ENTOMOLOGY & DEATH: A PROCEDURAL GUIDE

Edited by -

E. Paul Catts
Washington State University

and

Neal H. Haskell
Purdue University

Illustrated by E. Paul Catts

Joyce's Print Shop, Inc., Clemson, South Carolina

Published in the United States of America by Joyce's Print Shop, Inc.
Mailing address: P.O. Drawer 1606, Clemson, SC 29633.

Library of Congress Cataloging in Publication Data

Catalog Card No.: 90-064360

E. Paul Catts, Neal H. Haskell

Entomology & Death, A Procedural Guide

Clemson, SC: Joyce's Print Shop, Inc.

ISBN 0-9628696-0-0

Dedication

This manual is dedicated firstly to our friend and colleague, Theodore R. Adkins, Jr., Ph.D. who died suddenly of a heart attack (August 1989) before this work was completed. Ted was an effective, outspoken and enthusiastic proponent of forensic entomology. We were all greatly saddened by his death and miss him very much.

Secondly we dedicate this work to those luckless victims of homicides whose plight has made this endeavor necessary, and to the unsung army of death investigators who devote their talent and energy to the onerous task of solving the most heinous of crimes.

Council of American Forensic Entomologists

— September 1990 —

Contents

Contents

Contents

Contents

Preface*

Even though forensic entomology soon will mark its centennial, until quite recently it has been ignored generally as a tool in forensic investigations. There appears now to be growing interest by law enforcement people and forensic scientists alike to bring entomology into the company of other proven technology in death investigations. This handbook reflects that growing interest.

This book had its beginning at an informal gathering of most of the authors appearing here during the 18th International Congress of Entomology in Vancouver, British Columbia, during the summer of 1988. They had presented a standing-room-only forensic entomology symposium during this gathering. After several years of trading experiences and mutual assistance in case studies, this group concluded that a practical, vernacular guide to the collecting and processing of the hordes of little creatures inhabiting a decaying corpse was needed. It is toward this goal that we offer this work. As may be anticipated, we have not been able to get away from scientific jargon completely, but we have tried hard to hold it to a minimum of essential terms.

The book is organized to carry you, the reader, from an introductory background in forensic entomology and its current status, through example case studies, and on to a brief introduction to the structure and biology of insects and other similar creatures associated with carrion. Then it proceeds to lead you through the standard death scene investigative procedures, the collection and field processing of entomological evidence, either at the scene or at the autopsy, and the laboratory processing and analysis of this evidence. Finally, it treats the record keeping that flows from this investigative process and the courtroom testimony and ethical considerations attendant to the legal process.

We have kept the physical dimensions of the book relatively small so that it may be carried to and used at the scene. We have also chosen to have it wire bound in leaflet page form so that it is easier to hold open where desired.

We are offering this book to a mixed audience of school-educated and scene-educated death investigators as well as to those "closet" forensic entomologists who are curious about practices and procedures in forensic sciences and police investigations. We hope it will fill the large gap that exists between the law officer at the scene, the coroner in the morgue, and the entomologist in the museum.

Collectively the authors have well over 100 years of experience in the study of insects in crime investigations. From the veteran "father" of American forensic entomology, Bernie Greenberg, to the most recent

* This document contains neither recommendations nor conclusions of the Federal Bureau of Investigation.

member of our happy band, Neal Haskell, each one sports a slightly different area of expertise in this young science. Most are associated with academic institutions as entomologists, although C. Andrews and Wayne Lord are with state and federal law enforcement agencies, respectively. For example, Lee Goff and Haskell have the greater experience in collecting entomological evidence at the death scene and the autopsy, while Greenberg has a longer record of courtroom testimony as an expert witness and Rob Hall's forte is in the taxonomy of insects, especially blow flies.

The logo appearing on the cover of this book also deserves comment. The human skull and hand magnifying lens represent the investigative nature of police work in violent crimes. Superimposed on this is a fly maggot and in the background the circular posterior spiracular plate of a blow fly. Both lend emphasis to the role of insects in death investigations and to blow flies in particular as conspicuous actors in corpse decomposition. We call ourselves the Council of American Forensic Entomologists - note the acronym CAFE - and this cover design is our (the council's) logo, too.

Even though a few photos may have been more effective in some parts of this work, we have sacrificed their realism in order to hold down publication costs by doing line illustrations throughout. In most cases, line drawings present a clearer, more concise picture anyway.

September 1990

E. Paul Catts
Neal H. Haskell

Acknowledgments

We, the editors and contributors, here express our sincere gratitude to the many people who have contributed in a diversity of ways to the development of this handbook.

We especially recognize the work of the late David G. Hall, Jr. in the taxonomy of blow flies and the continued contributions to this work by such U.S. National Museum scientists as Curtis W. Sabrosky and Raymond S. Gagne.

Many others have helped in this endeavor, each in his or her own way, and although it is not adequate to their efforts, we simply thank them here by listing them in order to award them some much deserved recognition. They are: Dennis Anslover, William C. Bass, Arthur Bohannon, Michele J. Catellier, Valerie J. Cervenka, Michael A. Clark, Ruth Clark, Paul V. Fournier, Allison Galloway, William D. Haglund, James and Patricia Hamby, B. Joel Haskell, Buthene Haskell, Dean A. Hawley, Alexandra Henson, Fred B. Jordan, Doris Lohrey-Birch, Murray K. Marks, Randall Pedigo, John E. Pless, Louis N. Sorkin, Ron Steverson, Gwendolyn Stokes, Cheryl Towell, William J. Turner, John W. Hicks, Jack Reitzell, Christopher Fiedler, James A. Wright, Arthur E. Westueer, Jr., Debra H. Lord, William C. Rodriguez III, John E. Douglas, Judson M. Ray, H. Wayne Caruer, Beverly DeCato, Edward T. McDonough, James L. Luke, Henry C. Lee, William B. Grothaus, Brian W. O'Keefe, Richard S. Farley, Charles S. Morris, Alvin I. Omori and Charles B. Odom.

We editors also accept full responsibility for any errors of omission or of fact in this edition. We welcome your suggestions, criticisms and comments for improvements.

Contributors

C. S. Andrews
Louisiana State Police
Baton Rouge, LA 70803

E. Paul Catts
Department of Entomology
Washington State University
Pullman, WA 99164-6432

M. Lee Goff
Department of Entomology
University of Hawaii
Honolulu, HI 96822

Bernard Greenberg
Department of Biological Sciences
Univ. of Illinois, Chicago Circle
Chicago, IL 60690

Robert D. Hall
Department of Entomology
University of Missouri
Columbia, MO 65211

Neal H. Haskell
Department of Entomology
Purdue University
West Lafayette, IN 47907

Wayne D. Lord
Laboratory Division
Federal Bureau of Investigation
Washington, D C 20307

C. Lamar Meek
Department of Entomology
Louisiana State University
Baton Rouge, LA 70803

Ralph E. Williams
Department of Entomology
Purdue University
West Lafayette, IN 47907

CHAPTER 1

MEDICOCRIMINAL ENTOMOLOGY

Robert D. Hall
University of Missouri, Columbia, MO

The typical death-scene investigator learns quickly that maggots and corpses go together. For many years, the "worms" crawling in the eyes, nose, and other orifices and wounds on dead bodies were considered just another disgusting element of decay — something to be rinsed away as soon as the corpse was placed on the table for autopsy. While ballistics, firearm examination, bitemarks, gunpowder residue chemistry, blood spatter analysis, and other elements of scientific criminology were studied and refined, the insects associated with death scenes were largely ignored. Through the years, however, a few scientists have researched forensic entomology, which has become a fascinating, and at the same time a more arcane, field of biological study. The potential for contributions of entomology to legal investigations has been known for at least 700 years, but only within the last decade or so has entomology been defined as a discrete field of forensic science.

Forensic entomology is the name given to any aspect of the study of insects and their arthropod counterparts that interact with legal matters. Lord and Stevenson (1986) divided it into three components: **urban** entomology (legal proceedings involving insects and related animals that affect man-made structures and other aspects of the human environment), **stored products** entomology (proceedings involving insects infesting stored commodities such as cereals and other kitchen products), and **medicolegal** entomology. The latter field, sometimes termed "forensic medical entomology," and in reality "medicocriminal entomology" (because of its focus on violent crime), relates primarily to 1) determination of the time (postmortem interval or PMI) or site of human death, 2) cases involving possible sudden death, 3) traffic accidents with no immediately obvious cause, and 4) possible criminal misuse of insects

(Leclercq 1969). Forensic entomology is inextricably linked with the broader scientific fields of medical entomology, taxonomy, and forensic pathology.

SCOPE OF MEDICOCRIMINAL ENTOMOLOGY

As presently practiced, medicocriminal entomology deals mainly with inferences made after examination and identification of arthropods collected from or near corpses. Forensically significant conclusions often can be drawn by noting the state of successive colonization of a corpse by local arthropod fauna, or by identifying the developmental stage of necrophilous (dead-flesh eating) insects collected in, on, or near the body. A professional entomologist is skilled in obtaining an adequate sample of appropriate arthropod specimens, properly preserved for study, from a death scene. In most cases, additional specimens need to be reared to the adult stage, and a professional entomologist knows how to accomplish this successfully. Knowledge of arthropod (especially blow fly) biology and geographic distribution can allow the determination of accurate estimates relative to the interval of time that a body has been exposed to arthropod activity, and an indication of whether the fauna collected is indigenous or foreign to the site where the body was found. Often this can prove useful in determining the actual location of death. For example, certain species of blow flies tend to be found primarily within large urban centers. Identification of such species in association with a corpse found along a rural roadside suggests that the victim was killed in town and subsequently dumped in the remote rural environment.

The ability to identify arthropods accurately is the principal role played by medicocriminal entomologists. This more than any other factor provides the solid foundation for all subsequent inferences relative to that fauna. Although specimens may be collected by crime scene technicians, medical examiners, or forensic pathologists, skill in identification probably will rest with a fully qualified entomologist.

Case histories have documented the utility of medicocriminal entomology and point out the unique contributions of this field of science. Nuorteva et al. (1967, 1974) presented a series of cases from Finland in which blow flies were used as indicators for indoor as well as outdoor death scenes, and where the immature (larval) or adult stages were used for identification. Leclercq (1969) provided a typical case scenario and outlined how insect data were used to corroborate information obtained from other sources. Greenberg (1985) outlined several cases, including a description of how laboratory fly-rearing data were used to calculate the number of accumulated degree-hours (ADH) required for certain blow fly species to develop, and how such data were applied to the solution of a murder case in Illinois. In another recent case, Greenberg (pers. comm.) described how the absence of insects, in a seemingly straightforward death scene, led to a killer's confession. A window next to the victim was open when the body was found, thus giving the impression that the murderer had forced entry into the room the night before. However, the air conditioned room was cool even though it was very hot outdoors. In reality, the killer was known to the victim, had a key, and had returned to "set the stage" by opening the window just prior to feigning discovery of the corpse. The insects thus had insufficient time to colonize the body because the window had been closed prior to the return of the killer. When confronted by this biological reality as pointed out by medicocriminal entomology, the killer confessed.

HISTORICAL BACKGROUND OF MEDICOCRIMINAL ENTOMOLOGY

In an experiment famous as much for its demonstration of scientific method as for its contribution to entomology, Francesco Redi (1668) studied rotting meat that was either exposed to or protected from flies. From his analysis of subsequent blow fly infestation, he refuted the hypothesis of the "spontaneous generation" of life. Up to that time, it was generally believed that under the right conditions maggots came from rotten meat. Later, Bergeret (1855), near Paris, France, was the first westerner to use insects as forensic indicators. The body of a baby was found behind the plaster mantle in a house, and an investigation was begun. Bergeret determined that the assemblage of insects associated with the corpse pointed to a state of decay that dated back several years; consequently, the question of guilt was thrown upon the earlier occupants of the house, and not upon the current ones. Bergeret's methods and materials were quite similar to one of the main medicocriminal entomological techniques still in use today, that is, the successive colonization of a corpse by a predictable succession of arthropod species. Between 1883 and 1898, J. P. Megnin in France published a series of articles dealing with medicocriminal entomology. The most famous of these, *La Faune des Cadavres: Application l'entomologie a la Medicine Legale* (Megnin 1894), served in large part to make the medical and legal professions aware that entomological data could prove useful in forensic investigations.

Although entomologists are most familiar with the references cited above, medicocriminal matters in the Far East predate these considerably. In 1235 A.D., Sung Tz'u, a Chinese "death investigator," wrote a book entitled *The Washing Away of Wrongs* (as translated by McKnight 1981) in which forensic science as known at that time was detailed. In this text, what was probably the first actual medicocriminal entomology case was recounted. A murder by slashing occurred in a Chinese village, and the local death investigator was deputized to solve the crime. After some fruitless questioning, the investigator had all villagers bring their sickles to one spot and lay them out before the crowd. Flies were attracted to one of the sickles, probably because of invisible remnants of tissue still adhering to it, and the owner subsequently broke down and confessed to the crime. In other portions of the text, Sung Tz'u demonstrated knowledge of blow fly activity on bodies relative to those orifices infested, the time of such infestation, and the effect of trauma on attractiveness of tissue to such insects.

Any analytical system is as reliable as is the data upon which it is founded, and forensic entomology is no exception. Because accurate identification of necrophilous arthropods is of paramount importance, few repeatable results could be obtained before adequate taxonomic work had been accomplished on the invertebrates (the insects and related animals) in question. Taxonomy and systematics comprise the science describing, classifying, and proposing evolutionary relationships of the various forms of life. Although many synanthropic (strongly associated with human activity) flies (such as *Drosophila, Musca, Muscina, Ophyra, Stomoxys,* and others) are not encountered frequently in typical forensic investigations, other species assume great importance (Greenberg 1985). Carrion (dead tissue)-feeding blow flies (Calliphoridae) and flesh flies (Sarcophagidae) are those most useful in death investigations. Aldrich's (1916) monograph on the Sarcophagidae made use of distinctive male genitalia, thereby enabling entomologists to identify adult male specimens from this important family. This concept involved the so-called "lock-and-key" arrangement in many insects that facilitate reproductive isolation between species. The male copulatory organs of each kind (species) of higher flies are composed of unique, complex structures that are used as key characters to enable specific determination. This adaptation has been applied with equal success to the forensically-important blow flies.

Twenty years later, Knipling (1936) published descriptions and keys to many common early (1st instar) maggots of flesh flies. Although considerable work had been done on the blow fly fauna of North America (for instance, Knipling 1939), Hall's 1948 monograph, *The Blowflies of North America,* made possible the accurate identification of adults and mature larvae of most species of this family as well. Although very few new (that is, previously unrecognized) North American calliphorid species have been described recently, efforts have been devoted to accumulate improved distributional information (Hall and Townsend 1977, Hall 1979, Goddard and Lago 1983). More research is needed on accurate identification of the critical larval and pupal stages (those most frequently collected in death investigations). At present, first instar blow fly larvae (the stage that hatches directly from the egg) generally are not identifiable to species, and second instars (the next maggot stage) can be identified accurately only on occasion. The situation is somewhat better with respect to third instar or prepupal larvae (the largest maggot stage, and that most commonly observed), but only if such specimens are preserved properly. Even so, a significant number of indigenous blow flies cannot be identified at present as immatures. This is currently an area of active research, and to this end the relatively new technique of scanning electron microscopy is being applied (Liu and Greenberg 1989).

Because of the medicocriminal requirement for reliable data on rates of larval development, considerable effort has been expended to measure such intervals. Anecdotal information on blow flies contained in earlier works was largely supplanted by Hall's (1948) rearing data, and the latter has been refined for some forensically-important species to degree-hour status (Greenberg 1985). Because insects are cold-blooded animals, their rate of development is more or less dependent on ambient temperature. Research has shown that for each species there generally is a threshold temperature below which no development takes place. As temperature rises above this threshold, a certain amount of time is required for the insect to attain defined stages of development (for instance, from the newly-laid egg through the second instar maggot). Because this heat is accumulated as "thermal units," it can be calibrated and described as "degree-days" or "degree-hours," depending on the accuracy of temperature readings and time period involved. However, most laboratory rearings (upon which the degree-hour data are developed) have been done at constant temperature, so additional research will be necessary to establish correlations between these data, typical fluctuating field temperatures (warmer during the day and cooler at night), and the average daily measurements frequently reported from weather stations. Retrospective weather records from the nearest weather-recording station (such as an airport) are those most often used in medicocriminal evaluations.

Access to the scientific literature pertaining directly to medicocriminal entomology has been facilitated by two recent bibliographies. An initial guide to entomological involvement in forensic pathology, plus a selected bibliography, was provided by Meek et al. (1983). A bibliography of all publications dealing wholly or in large part with medicocriminal entomology worldwide was compiled by Vincent et al. (1985). The latter paper contained 329 references and was current through 1983; therefore, the actual body of literature pertaining to this subdiscipline of forensic entomology is not large when compared to many other biological or legal subjects. The first textbook devoted to forensic entomology was published in 1986: *A Manual of Forensic Entomology* (Smith 1986). This is an excellent reference for the entomologist, and it brings together in one place all the salient information contained in the literature on this subject.

PROFESSIONAL STATUS OF MEDICOCRIMINAL ENTOMOLOGY

The most current worldwide directory of forensic entomologists (Lord and Stevenson 1986) listed a total of 62 scientists involved in this field of study. Of these, 33 (53 percent) were linked solely with the "medicolegal" subdiscipline, and an additional five (eight percent listed "medicolegal" entomology as one of their forensic specialties). Most (44 percent of the total) were affiliated with colleges or universities, taught entomology or biology and performed research or worked with the Cooperative Extension Service. A few conducted research primarily on medicocriminal entomological subjects. A significant number (35 percent) functioned as consultants — working for private industry, being self-employed, or having retired from previous employment in entomology. U.S. medicocriminal entomologists who were members of the Entomological Society of America (the scientific "society" of entomologists in the U.S.) were generally affiliated with Section D (Medical and Veterinary Entomology) of that organization. Only about 39 percent of such U.S. workers were members of the American Registry of Professional Entomologists (ARPE), the formal organization comprised of professional entomologists, and still fewer are members of the American Academy of Forensic Sciences. Following the lead of physicians (who draw their professional identity from the American Medical Association), attorneys (who have the American Bar Association), and other professionals, it will become increasingly necessary for medicocriminal entomologists to affiliate through a professional organization with strict educational, ethical and maintenance standards. Facilitation of qualification as expert witnesses is one reason for encouraging such a connection.

Most (82 percent) current medicocriminal entomologists hold an earned Ph.D. degree or its equivalent (Lord and Stevenson 1986). The remainder possess M.S. or M.D. degrees. Given the requirement for demonstrating "expertness' as part of the medicocriminal procedure, it is likely that an earned Ph.D. with emphasis in cogent areas of medical entomology, taxonomy and allied disciplines will remain essential. The possibility of combining graduate entomological education in the medicocriminal field with ancillary professional medical or legal degrees remains largely unexploited. One compelling reason for this may be that, at present, few entomologists worldwide are employed full-time as medicocriminal experts. In at least one instance, an individual with prior training in criminology entered a graduate program with emphasis on medicocriminal entomology.

Although the situation may be changing slowly, medicocriminal entomologists are seldom included on routine crime-scene investigation teams. Whether this has occurred because such specialists are rare, because their presence is inconvenient or impractical, or because their area of expertise is not appreciated fully for either its uniqueness or potential contribution to forensic pathology is immaterial. A major result has been that an important responsibility of medicocriminal entomologists consists of training crime-scene technicians and allied personnel to recognize, obtain, and properly preserve entomological evidence. The entomologist then reviews such evidence and often provides expert assistance if entomological data appear crucial to the case. Important to the latter function are eventual qualification as an expert witness, the increased professional status of entomologists toward that end, a facility with courtroom protocol, and a fee structure related to this effort.

Most up-to-date investigative units acknowledge in their standard operating procedures (SOP's) the potential for acquisition of entomological evidence. The purpose of this Procedural Guide is to facilitate that objective.

CHAPTER 2

CASE HISTORIES OF THE USE OF INSECTS IN INVESTIGATIONS

Wayne D. Lord

Federal Bureau of Investigation, Washington, DC

Within the United States and around the world, insect scientists (entomologists) are being called upon with increasing frequency to apply their knowledge and expertise to criminal and civil proceedings and to become recognized members of forensic laboratories and medical/legal investigation teams. Insect evidence gathered from and around the corpse, when properly collected, preserved and analyzed by an experienced and appropriately trained forensic entomologist, can provide an objective estimate of the time of death as well as other valuable information concerning the circumstances surrounding the victim's demise.

INSECTS AND HUMAN DECOMPOSITION

Human corpses, whether they have been produced naturally or as the result of foul play, are processed by insect decomposers in the same manner as any other piece of carrion. Forensic entomology, therefore, is based on the analysis of the insects and other invertebrates which sequentially colonize a corpse as decomposition progresses and on the rates at which the various stages of their offspring develop. Entomological information can be extremely useful in determining manner of death, movement of the corpse from one site to another, and length of the postmortem interval. The following case history studies are described to illustrate this use.

While a wide variety of insect species are attracted to decomposing remains and play an active role in the decay process, two groups, the flies (Diptera) and beetles (Coleoptera), are of major importance in most circumstances. Diptera, whose larvae are capable of living in a semi-liquid medium, are the first insects to be attracted to and to colonize decomposing remains. Fly larvae (maggots) are responsible for the dramatic consumption of the corpse's tissues. Only much later, when the corpse has to a large extent dried out, do species of other insect groups, notably beetles, move in and continue the process.

Case No. 1

On 4 June the partially clad body of a young female was found alongside a rural highway in the northwestern United States. An autopsy revealed that she had died of multiple head and neck wounds inflicted by a heavy sharp object. She was subsequently identified as a 14-year old prostitute. Her brother reported her as missing approximately four days prior to discovery of her corpse.

She had last been seen alive on the morning of 31 May in the company of a 30-year old army sergeant, the primary suspect. While considerable circumstantial evidence supported the theory that the victim had been murdered by the sergeant, an accurate estimation of time of death was crucial to establishing a possible link between the suspect and the victim at the time when death occurred.

Several estimates of postmortem interval were offered by medical examiners and investigators. These were based largely on the physical appearance of the body and the extent to which decompositional changes had occurred in various organs. They were not based on any quantitative scientific methodology.

Numerous fly larvae (maggots), adult flies, and other insects were observed and collected in and around the victim's wounds. Some were placed alive in small containers and subsequently reared to produce adult flies. Others were placed immediately into a liquid preservative. Additional specimens collected at the autopsy were processed in a similar manner. Numerous photographs of the crime scene, the surrounding vegetation and terrain, and the corpse were taken. These photographs included enlargements illustrating the adult flies and maggots present at the time the body was discovered.

Reports describing the condition of the body when found and detailing autopsy procedures and results also were reviewed. Climatological data, including maximum and minimum temperatures, incidence of rainfall, cloud cover, wind speed and direction, and relative humidity, were obtained from a government weather station located a short distance from where the victim was found. These data indicated the environmental conditions to which the remains and its insect associates were exposed.

Based on this total array of evidence, entomologists determined that the first insects to colonize the remains had arrived on 31 May. The insect evidence indicated a PMI of four days.

Based on this evidence, the army sergeant with whom the victim had been last seen alive was arrested and charged with first degree murder. On questioning he admitted to having murdered the victim by striking her six to eight times with a small hatchet sometime around noon on May 31. Subsequently, he entered a plea of guilty to the murder charge and was sentenced to life in prison without parole.

THE IMPORTANCE OF BLOW FLIES

Fly species differ in abundance from region to region, from habitat to habitat, and from season to season. In the northern United States for example, a blue bottle fly, *Calliphora vicina*, is more abundant during the cooler parts of the year, whereas a green bottle fly, *Phaenicia sericata,* dominates corpses during the warmest parts of the summer. Another green bottle fly, *Lucilia illustris,* frequents corpses located in open, brightly lit habitats, whereas the black blow fly, *Phormia regina,* prefers shade.

Case No. 2

The remains of a murder victim were reportedly thrown down an open well on a small farm in a rural area in south-central Indiana. Then the well was completely filled with junk, tires, and rocks. The exact location of the well where the remains were deposited was unknown, but as the investigators drove into one of several wooded farm yard sites being investigated, it was obvious they had found the right location. Several thousand flies were hovering over a pile of old tires. The remains were found at the bottom of the well under the debris. Decomposition was advancing in the body, but there were no insects found on it. Access to the body by the blow flies was prevented by the intervening material, but odors were still capable of attracting multitudes of insects.

Case No. 3

Recently, in New England, the fully-clothed body of a young adult woman was found in a parking lot located behind an urban industrial complex. The victim had died from a single, 9mm bullet wound to the right temple, and a substantial pooling of blood was noted beneath the victim's head. The body was discovered initially by employees reporting for work at approximately 6:00 a.m. No insect evidence was observed on or around the body during the preliminary crime scene survey. As the investigation proceeded, however, and the body was warmed by the morning sun, small numbers of green bottle flies *(Phaenicia sericata)* were observed feeding at the bullet wound. By the time the corpse was removed from the scene, patches of eggs were present in and around the wound. Knowing that the previous day's climatic conditions were ideal for blow fly activity, that adult blow flies are not typically active at night, and that blow flies would be highly attracted to the pooled blood, entomologists concluded that the victim had been killed during the hours of darkness preceding the discovery of the body. It was later determined that the young woman had been last seen alive around midnight of the previous day.

THE IMPORTANCE OF BLOW FLIES: EGGS

Blow fly eggs, collected from remains and analyzed, can provide investigators with an accurate estimate of postmortem interval and allow them to focus more narrowly their investigative efforts.

Blow fly eggs are small (2 to 3 mm), whitish-yellow, and somewhat elongate. During warmer seasons they are frequently packed into natural body openings and wound sites and in large numbers are easily visible to the naked eye. In temperate areas during colder months, however, their numbers may be few, and they may be difficult to locate, being more deeply recessed in cryptic locations. Blow fly eggs will typically hatch within one to three days depending on species and environmental conditions. Dissection of egg samples and analysis of the state of embryonic development may further delineate the time since oviposition and, therefore, the time of the victim's death.

Case No. 4

On a midmorning in August, the half-nude body of a young female was discovered, more or less face down, among a group of junk automobiles near Spokane, Washington. The victim had died of multiple stab wounds to the chest and neck, and adult blow flies were observed in and around the wounds. Blow fly eggs were collected from the wounds at autopsy in the late afternoon. Subsequent dissection of the eggs showed no embryonic development suggesting that they had been deposited on the remains less than eight hours earlier.

The victim had been last seen alive during the evening two days prior to her discovery. The insect evidence, however, suggested that the young woman had been murdered during the hours of darkness preceding the finding of her remains. Had the victim died any earlier, young fly larvae (maggots) rather than eggs would have been collected from her wounds. Climatic conditions on both days prior to her discovery were suitable for adult blow fly activity and egg laying. A subsequent investigation verified these findings, revealing that the victim had been murdered during the hours of darkness just prior to the morning of her discovery.

Case No. 5

In early Spring, the fully-clothed body of a young, white male was found in a sandy, shrub habitat, in the southwestern United States. The victim had died of multiple small caliber bullet wounds to the chest and back. There was little external evidence of decomposition. A small amount of blood was observed to have oozed from the victim's left nostril, partially coating his left eye.

A small irregular granular mass was collected from the surface of the victim's left eye. Closer examination of this substance revealed it to be a small mass of blow fly eggs. Several eggs had hatched already.

These maggots were reared to adults and were subsequently identified as *Cochliomyia macellaria,* the secondary screw worm. Based on the climatic conditions and knowledge of the developmental biology of this fly, it was determined that the eggs most likely had been laid on the corpse 24 to 36 hours prior to the time the corpse was found. Subsequent investigation determined the identity of the victim. He had last been seen alive in the company of a male companion approximately 36 hours prior to the time his corpse was found. It was later determined that the victim had been murdered by his companion approximately 36 hours prior to discovery of the remains.

THE IMPORTANCE OF BLOW FLIES: MAGGOTS

Maggots grow rapidly, passing through three growth steps (instars) punctuated by moults before becoming fully grown. Typically, large numbers of larvae hatch together and move around the corpse as a group. In doing so, they disseminate bacteria and secrete enzymes which enable them to consume virtually all of the soft tissues of the corpse. Their growth takes several days to several weeks depending upon species, environmental conditions, and the number of larvae present.

Case No. 6

In November a partially clad woman's body was found by passing motorists in a wooded area approximately 50 feet from a well-travelled highway in suburban Washington, DC. The deceased was a young, adult black female, weighing approximately 116 pounds. Her body was cold, rigor mortis was absent, and only minor external decompositional changes were evident. The victim had died of multiple stab wounds to the chest and neck. During the autopsy, several large maggots were observed migrating away from the corpse. This behavior is characteristic of fully developed, post-feeding blow fly larvae. Several additional maggots were removed from the neck wounds and clothing.

The victim was subsequently identified as a 21 year-old mother who resided with her infant and parents in an urban apartment complex in southeast Washington, DC. She had been reported missing by her parents 18 days prior to the discovery of her body. She was last seen alive during the early morning hours on Wednesday, 26 October in the apartment of the primary suspect, who had a history of sex-related offenses and violent crimes and had recently been released from prison.

Witnesses recalled having seen a man and woman struggling and having heard a woman screaming in the vicinity of the apartment house during the early morning hours 18 days prior to her discovery. Hair from the victim's head and pubes was found on bed sheets in the suspect's apartment. A shoe belonging to the deceased was subsequently found in a secluded urban woodlot located a short distance away. Samples of blood and hair taken from the suspect's car also were identified as belonging to the victim. An analysis of auto carpet fibers, removed from the victim's clothing, further demonstrated that she had been transported in the same vehicle.

While considerable circumstantial evidence pointed to the suspect, an accurate estimate of the time of death was critical in establishing the sequence of events surrounding the death. Several conflicting estimates of postmortem interval, varying from 2 to 8 days, were offered by medical examiners and case investigators, based primarily on the physical appearance of the body and the extent to which autolytic changes had occurred within various organs.

The maggots observed in and around the body during the autopsy were identified. Rearing of maggots to the adult stage was not possible as none of the specimens had been maintained alive. Soil samples yielded no additional specimens. Numerous photographs depicting the crime scene, the surrounding vegetation and terrain, and the body itself were examined.

Climatological data, including maximum and minimum temperature, cloud cover, rainfall, wind speed and direction, and relative humidity were obtained from a National Weather Service (NWS) observatory located less than a quarter of a mile from where the victim's body was found. Additionally, reports describing the condition of the body when found and detailing autopsy procedures and results were reviewed.

The largest fly larvae inhabiting the remains were fully engorged third instar larvae and post-feeding larvae of *Calliphora vicina*. One specimen showed morphological signs of pupariation.

Based on the temperature records for the relevant period, the number of days necessary for *Calliphora vicina* to develop from egg to prepupa was calculated. Because the average daily temperature to which the developing flies were exposed was cool (10C) and because two few larvae were present within the corpse to elevate microenvironmental temperature, the postmortem interval was estimated to be 15 days.

Based on the entomological data and the other evidence acquired during the police investigation, the suspect was arrested and charged with first degree murder, kidnapping, and felony rape. He was eventually tried, found guilty of all charges, and was sentenced to a lengthy prison term. Later it was learned that he had murdered the victim during the early morning 18 days prior to discovery and had hidden the body, by covering it with tree branches, a mattress, and other debris, in a nearby urban woodlot. Three days later, again during the early morning hours, he had transported the corpse to the location where it was found. The materials used to hide the body may have prevented oviposition by adult *Calliphora vicina* during the first few days following death. The fifteen day postmortem interval was valid.

Case No. 7

The semi-nude remains of a white male were discovered under a bridge in southern Indiana during the second week of December. The remains showed little decomposition, with only slight skin slippage on the arms and upper torso. A considerable portion of the decedent's face (left side) was skeletonized, postmortem damage caused by rodent feeding. A few calliphorid larvae (*Calliphora vomitoria)* were found in the exposed sinus and oral cavities of the face.

The remains appeared to have been exposed only a relatively short period of time based upon the physical appearance. In this case, cool temperatures associated with the particular season of the year were important in estimating the approximate time of death. For southern Indiana, temperatures were below average in October, above average in November, and above average in December. Analysis of the blow fly larva development and comparison with climatological data suggested that it would take at least 48 days at the recorded daily maximum and minimum temperatures for the maggots to reach the size and stage found on the remains. When the investigation was completed and the remains identified, it was found that the decedent had wandered away from a mental institution 54 days prior to the discovery of the corpse. The decedent had died from hypothermia and exposure during an October night when temperatures fell well below freezing (24F).

Case No. 8

In the southeastern United States during mid-November, police were called to investigate a foul smelling odor which was emanating from a small single family home in an impoverished section of town. Investigating officers soon discovered the badly decomposed body of a young, black female in a shallow grave in the dirt basement of the dwelling.

The victim had died of a single bullet wound to the head inflicted with a small caliber rifle. Subsequently, a careful examination of the corpse and a detailed excavation of the soil in and around the grave site revealed the presence of numerous larvae of *Calliphora vicina* and larvae and pupae of a relative of the house fly, *Synthesiomyia nudesita*.

Specimens collected from the scene were reared in the laboratory. Supplemental information including climatic data and soil temperatures were reviewed in an effort to determine the intervening climatic conditions. Using information on the developmental biology of both of these species of flies, an estimate was made that the victim had died and was colonized by flies approximately 28 days prior to the time of discovery.

Investigators were able to target their investigation in and around the estimated time of death. Shortly thereafter a suspect was identified. This individual eventually confessed to having killed the victim 28 days prior to the time the body was found. She had attempted to bury the victim in a shallow grave located in the basement of the house shortly after committing the homicide.

In this case, larvae of multiple species of flies provided investigators with the only scientifically reliable method of estimating the time of the victim's death.

Case No. 9

Similarly, in early summer, the body of a young, unidentified, white female was found at the end of a logging road in a rural section of the northeastern United States. Investigation revealed that the unidentified young woman had died from a single shotgun blast to the right side of the head. While processing the crime scene, investigators collected representative samples of blow fly adults and maggots infesting the gaping wound.

These specimens proved to be mature larvae and adults of the black blow fly (*Phormia regina*). Following an indepth review of climatic conditions, habitat characteristics, and crime scene photographs, entomologists estimated that the young woman had been killed about five days prior to her discovery and that her body had been placed in that location close to the time of her death.

Based in part on the entomological analysis, investigators circulated requests for reports of missing persons fitting the deceased's description who had disappeared five days prior to the corpse discovery. Eventually, the young woman was identified and her boyfriend became the prime suspect. Just prior to his arrest, however, he was found hanged in a motel. A suicide note revealed that he had murdered the victim five days before her remains were found and had committed the homicide at the end of the logging road where she was discovered.

Case No. 10

On 26 August, a 37-year old black male was found in Kawainui Marsh, a swamp located approximately 1.6 km from the ocean on the windward side of the island of Oahu, Hawaii. The elevation for the site was only slightly above sea level. The body, clothed in blue jeans and a t-shirt, had a gunshot wound to the right frontal skull.

The corpse was in an advanced state of decay. The skull was largely devoid of flesh, although some was present on the lateral surfaces and the ears were relatively intact. The cervical area was largely devoid of flesh and the vertebrae were clearly visible. The groin area was largely decomposed and contained a large number of Diptera larvae. On autopsy, no penetration of the abdominal cavity by larvae was observed. Collections of insects from the remains yielded 2 species of Calliphoridae larvae. *Chrysomya megacephala* was represented by 3rd instar larvae from the chest, groin, and legs, while *Chrysomya rufifacies* was represented by 2nd instar larvae and 3rd instar larvae from the chest. None of the larvae from either species were in the prepuparial stage.

Temperature data were obtained from the nearest weather station, and climatic information from this station was analyzed for the time period in question. This analysis, when coupled with the developmental stages of the specimens collected, suggested a minimum postmortem interval of approximately 117 hours.

A subsequent investigation revealed that the decedent had last been seen at about 6:00 p.m. on 21 August when he departed for work. He failed to arrive at work as scheduled at 8:00 p.m. His remains were discovered on 26 August and were placed into a refrigerated crypt. This species *C. rufifacies* does not form dense maggot masses, activity for this species would have ceased shortly after the remains were placed into the refrigerated crypt. The postmortem interval based on insect age in this case was about 117 hours and fit well with the last sighting of the decedent, and his failure to report for work.

Case No. 11

The nude body of a 16 year old caucasian female was discovered in a road-side ditch about one mile from SEATAC Airport at 1000 on 31 August. The deceased was positioned face-down in an open fetal position with complete exposure to sunlight on a gently sloping berm of the roadside ditch. The lower body, up to the buttocks, was in a black plastic garbage bag. Maggots were found infesting the vaginal, oral and nasal openings and samples were collected for rearing and identification. Blow fly eggs were observed in the mouth and nostrils on 31 August and hatched the next day. Rigor mortis appeared to be lacking, but bloat, green belly discoloration and some tissue marbling was noted.

Weather at nearby SEATAC Airport had been warm and generally sunny over the past several days with temperature highs ranging from 73F (28 August) to 92F (31 August). The core body temperature at autopsy was 87F. Live maggots had been placed on whole beef flank tissue upon receipt and adult flies eventually emerged. The larger maggots were identified as *Lucillia illustris,* and maggot rearing yielded only *Phormia regina* adults. The larger maggots were from the vagina, whereas those that were reared had been collected from the mouth and nostrils.

Following a detailed analysis of the conditions to which the developing maggots were exposed, entomologists concluded that the victim had been killed 2.5 - 3 days prior to the discovery of her body. Warm environmental temperatures, continued exposure of the body to direct sunlight, and the likelihood that the victim's lower body had been covered by a dark colored plastic bag fostered a more rapid development of the blow fly larvae feeding on the remains than would normally have been expected. A subsequent investigation revealed that the victim had died during the night on 28 August.

Case No. 12

The decomposed body of an unidentified female was found in a conifer forest adjacent to a rural highway near Toledo, Washington in early May. Blow fly larval activity was conspicuous. The victim's right leg had been consumed to the bone up to the thigh, and the head, neck, and thorax were in an advanced state of decay. The abdomen contained the bulk of the maggot mass and adipocere deposits were observed on the lower back and buttocks.

Samples of blow fly maggots collected from the corpse were identified as fully developed *Calliphora vicina.* Under optimal conditions, development to this stage would require approximately 8 - 10 days; however, under cool, fluctuating temperatures, the developmental period could be expected to be twice as long.

The fact that these blow flies attacked the head and genital/anal regions of the deceased is not unexpected. On a fresh corpse, even one that is nude, blow fly attack will concentrate on the natural body openings, particularly the mouth and nose. Any added lacerations or battering of the head and neck enhances this behavior.

In this case, surface weather data from the Toledo Airport was an important key in determining the time of exposure of the deceased. Although *Calliphora vicina* adults have a flight activity threshold at temperatures lower than most other blow flies, they still require a span of several hours of sunny weather with temperatures above 55-60 F. The large number of maggots on the corpse and the apparently similar age of all maggots examined suggested that the corpse was exposed to a considerable number of gravid female blow flies during the early postmortem period. Close examination of surface weather data for the month preceding the discovery of the corpse showed three suitable windows of good "fly conditions" 1-2 April, 6 - 15 April, and 30 April - 2 May. The early May window was too recent to allow for advanced maggot development. The early April window was only marginally warm enough for fly activity and probably would not permit a large number of flies to lay eggs. The mid-April window was ideal, especially on 9 April when air temperatures reached 80F. A subsequent investigation verified the entomological findings.

Case No. 13

The body of a white adult male was discovered in low growth understory in a conifer woods at Ft. Lewis, Washington, 4 July. The deceased wore only a T-shirt, jeans and socks. The body was lying prone, face up, with the head partly beneath a small shrub at the edge of a clearing in the woods. The corpse was in late fresh stage of decomposition with an abundance of maggots on and in the upper trunk, cervical area and head. Scalp hair from the corpse had slipped as had some skin. The remaining facial skin was blackened and dried, and there was considerable fluid seepage into the soil beneath the anterior end of the body.

Three kinds of flies were identified in the material supplied. Predominant in numbers and biomass were 3rd instar maggots of the black blow fly, *Phormia regina*. This species is dominant in carrion during warmer summer months, particularly in shaded habitats. Its adults are often "early arrivals" at a corpse; gravid females may show up within 15 minutes of exposure. This species made up the vast majority of the maggot mass discovered during autopsy and accounted for nearly all tissue destruction.

The second fly type present were adult sphaerocerids (probably *Copromyza* sp.) These small black "gnats" first appear during early stages of bloat and represent the arrival of the second group (sere) of arthropods. Their presence indicates that the body was beginning bloat stage even though obvious manifestations were not noted.

The third type of fly was a muscid maggot, *Hydrotea* sp. This maggot is predacious on other fly larvae. Following a review of climatic data for the region and an assessment of the local habitat characteristics, it was estimated that the victim had died 3 to 4 days prior to the discovery of his remains. Warm July temperatures favored rapid larval growth even under the somewhat shaded conditions to which the body was exposed. The large population of maggots present and the advanced destruction of the cranial contents and neck tissues also suggested the possibility of the presence of perimortem head trauma. Subsequent investigation verified these estimates.

Case No. 14

Recently the fully-clothed body of a young adult female was found in a sparsely vegetated power line right of way in suburban Maryland. The young woman had left home for an evening walk several days prior and never returned. When found, the remains were in a moderate state of decomposition with large masses of maggots infesting the chest and neck. Additional maggot infestations were noted in the palms of the victim's hand. No insect specimens were collected at the crime scene or at the autopsy, and the death was ruled as a probable drug overdose.

Investigators eventually requested entomologists to review photographs of the remains to determine if the maggot age was consistent with the alleged time of the victim's disappearance. The location of masses of feeding maggots within the victim's chest, neck and palms clearly suggested the presence of perimortem trauma.

By the time this information was reviewed by authorities the victim had been buried. Based on the insect evidence, a court ordered exhumation was conducted. Careful examination of the skeletal remains by forensic anthropologists revealed evidence of stab wounds beneath the sites of maggot feeding. The death was then ruled a homicide.

In this case, atypical maggot infestations were the central indicators of traumatic injury.

Case No. 15

The bound, clothed body of an elderly caucasian woman was discovered in late fresh (pre-bloat) condition on 2 July at about 10:00 a.m. The corpse was lying on the ground in a densely shaded, understory and shrub stand of thimbleberry (*Rubus* sp.), ocean spray (*Holodiscus* sp.), and snowberry (*Symphoricarpus* sp.) A small wilted branch of the latter was squeezed between the third and fourth fingers of the victim's left hand. The hands were bound at the wrists, the head battered and the ankles bound with a rope that continued around the victim's neck from behind. Maggots of various sizes were noted, especially on the victim's face. A few adult flies were sitting on the body and face of the deceased during the midday removal of the body. A few were killed in alcohol for dissection and internal examination.

Live maggots completed their development (at room temperature, circa 24C) by 8 July. Adult flies began emerging on 14 July. A total of 20 flies had emerged by 16 July; all were killed and most pinned for examination. All adult flies proved to be *Cynomya cadaverina*. Preserved maggots were in three distinct size groups. The largest maggots were identified as 3rd instar *C. cadaverina.* Both medium and small sized maggots were 2nd instar *Phormia regina,* the black blow fly. Assuming that the maggot samples represented an approximate proportion of the size classes in the corpse, the largest maggots (*C. cadaverina*) were much lower in number.

Populations of *C. cadaverina* generally peak in late spring and early fall. The larger maggots represented about five days' development, meaning that the corpse was exposed to oviposition on 28 June and that the crime probably occurred prior to that date. This fly will frequent bodies in late fresh decomposition, so the body may have been transported for some distance prior to being dropped in the location where found.

The majority of the smaller maggots — those of medium and small sizes — were *Phormia regina* (the black blow fly). Eggs for these were laid on 30 June in large numbers and have an interpolated development time of 75 hours, or about 3 days. Rain on 29 June would have limited fly movement and allowed for greater fly activity on the following day, resulting in enhanced oviposition.

The entomological analysis, in this instance, suggested that the victim's body had been colonized by *C. cadaverina* on 28 June and secondary fly strike by the black blow fly, *P. regina,* occurred on 30 June. Because *C. cadaverina* will colonize remains which are somewhat decomposed, it was suggested that the victim may have met her demise several days earlier at a location different from where she was found. Other evidence at the death scene agreed with this conclusion independently.

Case No. 16

In July the partially clad remains of a white adult female were discovered in a landfill in suburban Maryland. The victim's body was badly decomposed with little soft tissue remaining, except on the extremities. The head, neck and chest were completely skeletalized, with the skull showing evidence of massive blunt trauma. Large populations of big maggots were recovered from the soil beneath the victim's head and torso. The body was located in a debris covered area which was devoid of vegetation other than short grasses and which exposed the remains to direct sunlight for most of the day.

Laboratory analysis of all specimens collected were black blow flies, *Phormia regina.* Although this fly is often an early arrival at remains, it also is more tolerant of older fresh bodies. Subsequent tests revealed that only a green bottle fly, *Phaenicia sericata,* was common at the landfill site where the body was found. Thus, it was suggested that the young woman's remains had not decomposed at the site where found, but rather that the body was moved from a different more shaded environment where it had been colonized by black blow flies.

The victim was later identified by investigators as a local prostitute who routinely provided services to truck drivers traveling along a nearby interstate highway. The young woman's body, it was surmised, had been kept for several days in a truck trailer until decompositional odors caused the driver to deposit the victim at the isolated landfill.

THE IMPORTANCE OF BLOW FLIES: PUPARIA

After reaching the third instar, fly larvae undergo a dramatic behavioral change during which they crawl away from the corpse, burrow down into the soil, and form a puparium containing the pupa. This process of pupation involves the "tanning" of the larva's outer skin, forming a casing around the body of the pupa. Within the puparium the adult develops and eventually emerges. Blow fly puparia are small, football-shaped objects, about the size of rat fecal droppings, and are reddish to dark brown in color.

Empty puparia are extremely durable and can remain in the soil beneath the corpse for hundreds of years. Blow fly puparia can supply valuable information long after the remains of the body have decomposed. Only by thorough examination of the area beneath the corpse or at a site where a corpse is reported to have decomposed will this evidence be found. When remains are located on substrates which are impervious to larvae burrowing, the puparia may be located in remote sites such as beneath bedding materials, boxes, within shag carpets, in drains, or attached directly to clothing.

Case No. 17

Unidentified human remains were discovered along a highway in west-central Indiana during mid-July. The corpse was fully clothed, but was completely skeletonized with only very small bits of soft tissue remaining. This suggests that the body had been in place at the scene for a considerable period of time. The fact that the decedent was wearing two pairs of trousers, an undershirt, flannel shirt, sweater, and jacket suggested that the person may have died during cooler weather, possibly during April or earlier.

Following the autopsy, it was concluded that the decedent had died from natural causes, but being unidentified, confirmation of identity was needed. Therefore, a search of the missing persons files was initiated for the time period during which the person was thought to have died. Originally it was thought, based solely on the physical appearance of the remains, that the individual had been dead for somewhere between 2 and 3 months or longer. Therefore, the investigation into the missing persons files would be initiated for the period (15 March - 15 May), thus consuming considerable effort, time and money.

Preliminary analysis of the entomological and climatological evidence suggested a much shorter time period for the postmortem interval. Viable puparia of the black blow fly (*Phormia regina*) were present indicating that the remains had been exposed for no more than 35 days. After completing the analysis, the forensic entomologist was reasonably sure that death had occurred about 30 days prior to the discovery of the remains.

When the case was finally resolved (the remains identified), it was found that the subject had last been seen hitchhiking in the vicinity of the death scene 31 days prior to the discovery of the remains. The entomological evidence proved very accurate and essential in pinpointing the specific time of death, and therefore aiding in the identification of the deceased.

Case No. 18

The remains of a 48 year old female in a moderately advanced stage of decomposition were found in a brush-filled ditch in a predominantly industrial area outside of Honolulu, Hawaii. The body was lying face up across the ditch, which was partially filled with water. The skull was largely devoid of flesh, and the mandible was separated from the skull. Three toes of the left foot were missing. The left hand was missing below the wrist, but the right hand was intact, although mummified. The remains were clothed in a blackened skirt. Upper, exposed portions of the remains were dry and partially mummified, while the lower, partially immersed, portions presented the appearance of the advanced decay stage of decomposition. Both feet were dehydrated, as were the right hand and arm. The rib cage was exposed, with some shreds of skin attached.

Further examination of the remains at the morgue revealed a largely skeletonized body lacking internal viscera and having only portions of parchment-like skin covering the neck, face, and lower extremities. With the exception of a fractured hyoid bone, no evidence of antemortem traumatic injury was found, although there was some evidence of postmortem vertebrate animal depredation. Due to the presence of the fractured hyoid bone, cause of death was determined to have been manual strangulation.

Collection of arthropods associated with the remains was made during autopsy. The portions of the body that were immersed in water yielded blow fly and flesh fly larvae. The former were identified as *Chrysomya megacephala,* and this identification was confirmed by rearing representative larvae to the adult stage. The flesh fly larvae were not successfully reared, and the species identity remains uncertain.

A greater variety of arthropods were present on the drier portions of the remains. Larvae of the Piophilidae, *Piophila casei,* were present in moderate numbers along with emergent 3rd instar larvae of a green bottle fly, *Phaenicia cuprina.* Empty puparial cases of another blow fly, *Chrysomya rufifacies,* were attached to the exposed ribs and leg bones. Adult dermestid beetles (*Dermestes maculatus*) were present on the external surfaces of the remains, and examination of the body cavities revealed a large number of both early and late instar larvae of this species. Clerid beetles, *Necrobia rufipes,* also were present on the external surfaces of the remains.

Analysis of the insect taxa and stages of development resulted in a postmortem interval estimate of 19.5 days. This estimate was based on the total assemblage of insects present on the remains and fit well with the last confirmed sighting of the decedent 20 days prior to the discovery of the remains.

The arthropods of primary significance in the postmortem interval estimate in this case were *C. rufifacies* and *D. maculatus,* as these were the species indicating the longest period of residence on the remains. On the island of Oahu, *C. rufifacies* requires 10 - 12 days for development under ambient temperatures. This species is one of the first to arrive at remains and oviposition continues for the first 4 - 5 days of decomposition. Presence of only empty puparial cases of this species indicated a minimum postmortem interval of 16 - 17 days. Dermestid beetles associated with remains in the Hawaiian Islands appear between days 6 - 8 in habitats similar to those of this case. Late instar larvae of *D. maculatus* in the body cavities were consistent with the 19 - 20 day postmortem interval. Also consistent with this time period were puparia of *P. cuprina* in the soil near the remains and the clerid *N. rifipes* on the drier portions of the remains.

The partial immersion in water of these remains served to keep them in a condition suitable for oviposition by adult flies longer than would be usual in a purely terrestrial environment.

Case No. 19

During November 1987, police investigators, responding to requests by neighbors and friends as to the whereabouts of a 69 year old male who had not been seen during the previous two days, discovered his corpse lying on the living room floor of a large home in an affluent neighborhood of a large Indiana city. There appeared to be no foul play involved, death by natural causes being established upon autopsy. Reports from neighbors also indicated that there were two elderly ladies (one an aunt, the other a sister of the male resident) in the residence who were being cared for by the deceased. The neighbors related that the ladies had been invalids (both bedridden) for years, and the brother/nephew had had many conversations with neighbors over his concern for these ladies' general welfare. As a neighbor stated to one of the investigating officers, "Just this week I asked him how the two ladies were getting along. He said, I don't know what I'm going to do with them, they kept me up all last night."

When the sheriff's deputies entered the bedrooms occupied by the sister and aunt, what the deputies found were two mummified corpses instead of the two elderly ladies needing help. One corpse was found in bed covered with bed clothes and women's clothing in a bedroom at one end of the home. The mummified remains had been slightly skeletonized by insect feeding. The extent of skeletonization was found to be primarily in the head and upper torso area, but the corpse was not completely skeletonized. Several hundred empty blow fly puparia were found. The second corpse was discovered in bed in another bedroom at the opposite end of the house and showed very little obvious skeletonization. However, the tissues were mummified and desiccated. Only two calliphorid puparia were found with the corpse. Also, a brown, stringy type material was found in large quantities in association with the soft tissues of both corpses. Upon analysis, this material was discovered to be fecal material from larvae of one of the dermestid beetles (*Dermestes lardarius*).

The lack of complete skeletonization by insect feeding indicated the possibility of establishing a seasonal window when death could have occurred. The corpse which had been partial skeletonization and contained the numerous puparia was estimated to have died in the fall (October), while the one without insect feeding was placed in early winter (late December to early January). While not being able to place the exact year of death using entomological evidence, it was concluded that both had been dead for a very long time, probably several years. These dates were later confirmed by physical evidence recovered at the scene. The woman, who, by the entomological evidence, was estimated to have died in October, had been keeping a diary. The last entry in her diary was on October 5, 1977, ten years earlier than the discovery of the remains. The other decedent had opened Christmas cards from 1977, but had not opened birthday and Easter cards from 1978. Upon autopsy, it was found that the elderly women had, most likely, died from natural causes. The diary of the aunt substantiated that both ladies were in poor health and failing, as she wrote.

Criminal involvement appeared minimal even though possession of unburied, unembalmed bodies in Indiana is only allowed by a burial permit or during an investigation through the Indiana coroner's system. Then it was discovered that social security and retirement checks from the women had been cashed and deposited by the brother/nephew over the intervening ten years (he had collected approximately $140,000). Having no close next-of-kin, a local church where the brother had been attending regularly had been named as the beneficiary of the estate. The retirement fund and the social security office demanded repayment of all funds which had been disbursed after death of the women. These funds would be obtained from the sale of property from the decendent's estate. In this case, the presence of the physical evidence, the corroborating entomological evidence, and handwriting analysis evidence supported the premise that both women had been dead for approximately the full ten year interval.

IMPORTANCE OF BEETLES AND OTHERS

Besides blow flies, a wide variety of other types of insects also colonize carrion. Rove beetles, carrion beetles, clerid beetles, sap beetles, checkered beetles, scarab beetles, and dermestid beetles all become members of the host corpse community feeding and rearing their young on the drier remaining tissues and on the large maggot mass. Additionally, other types of flies including house flies, flesh flies, skipper flies, fruit flies, and coffin flies also colonize the remaining tissues. The corpse eventually comes to support a complex and diverse community of insects often numbering hundreds of species and thousands of individuals. These insects, while far less numerous than the maggots which originally colonized the remains, can be seen flying, crawling, and scurrying about the corpse itself and in the soil beneath the remains.

These insects are attracted to the corpse in an orderly, progressive process termed succession. The successive nature of the colonization process enables the entomologists, when supplied with a representative sample of the insects present, to develop meaningful information concerning the circumstances surrounding the death of an individual. In a manner of speaking, by knowing the players on the field at the time that the corpse was found, forensic entomologists are able to determine the inning of the game.

Case No. 20

The remains of a female child (30 months of age) were recovered from a shallow grave on a narrow ledge on the side of Koko Head Crater, Oahu, Hawaii. The skeletonized remains were buried in dirt and gravel, and some bones were partially exposed. Other bones were scattered on the surface. The skull was facing upward, and the mandible was located several feet from the skull. Also present in the grave were four small stuffed dolls.

Examination of the remains at the morgue revealed largely skeletonized material. Present were the skull and mandible, most of the ribs, many thoracic vertebral bodies, the mostly skeletonized pelvis and lower extremities, the left humerus, radius, ulna, and scapula. Absent were many of the cervical vertebrae, right upper extremity, and scapula. Small bones of the left foot appeared to have been chewed and showed signs of postmortem animal depredation, as did the bones of the left forearm. Apparent scalp hair remaining adjacent to the skull was blond and straight, and measured up to 15 cm in length. Clothing accompanying the remains in which the corpse had been buried were a hooded jacket and a pair of running shoes. A second search of the dangerously precarious burial site the day following the recovery of the remains yielded the right scapula and arm as well as additional vertebrae.

Examination of the remains at the morgue yielded a limited assortment of arthropods. Empty puparial cases of the calliphorid *Chrysomya rufifacies* were attached to the skull under the scalp, which had been largely eaten away. Adult dermestid beetles, *Dermestes maculatus,* were present on the bones, and late instar larval skins of that species were observed in areas under the scalp, inside the cranial cavity, and on the femurs. These cast larval skins of *D. maculatus* were in good condition and did not appear to have been exposed to weathering effects for any period of time. Larval scenopinids (Diptera) were collected from the skull near the bases of the hair. Adult clerids, *Necrobia rufipes,* were recovered from the bones of the left foot. A silverfish (Thysanura: Leptismatidae) was recovered from the body bag.

In the hood of the jacket, there was a quantity of soil associated with the remains. This was processed in a Berlese-Tullgren funnel and sorted by hand. This sample yielded additional larval scenopinids and adult *D. maculatus.* There was an adult histerid beetle which represents an undescribed species. Mites associated with this sample included species in the families Acaridae, Histiostomatidae, Macrochelidae, Pachylaelapidae, Uropodidae, and Winterschmidtiidae.

Comparison of this assemblage with results of earlier decomposition studies conducted in xerophytic habitats on the island of Oahu resulted in a postmortem interval estimate of 52 - 76 days. This estimate was lowered to slightly over 52 days, based on the condition of the cast larval skins of *D. maculatus.* In the decomposition studies used for comparison, the last observed larval activity for *D. maculatus* was on day 51. The cast larval skins of *D. maculatus* disintegrate quite rapidly when exposed to weathering. The fact that these skins were in excellent condition and easily identifiable to species level indicated that they had been shed recently.

The window fly larvae (Scenopinidae) were all late instars and of a size comparable to those collected from decomposition studies on day 51 in one study and day 48 in another. This tended to reinforce the estimate of slightly over 52 days. The mites recovered also were consistent with this time frame, and the absence of certain other mite species was indicative of the shorter portion of the computer estimate. The final postmortem interval estimate of slightly over 52 days for this case fit well with the interval established independently by confession of the suspect in the case, the victim's father. Death proved to have occurred in the morning, 53 days prior to recovery of the remains.

Case No. 21

The skeletal remains of a 15½ year old caucasian female were discovered on the floor of a mixed oak/maple woodland in the Cumberland Mountains on 29 January. Weathering lines on the skull indicated that the occipital foramen was positioned to allow insects access to the cranial vault. Inside the cranium was a paper wasp nest of about 100 cells. All except 6-8 cells appeared empty, indicating that the full-sized nest had been constructed and occupied during the previous year. There was a loose electric cord close by suggesting that the deceased had been strangled. Also present was a weathered pair of denim jeans.

The wasp nest measured about 8 - 9 cm in diameter with cells of 5 - 6 mm diameter. The few cells retaining caps also proved to be empty. All signs indicated that a large active colony of *Polistes* wasps had spent the previous summer inside the cranium. *Polistes* probably begin their nest site searches in mid-April in this locality. Their sites must be protected and dry. This means that the cranium had to be cleaned of tissue and decay by early Spring of the previous year.

A single sphaerocerid puparial case was found in the cranium. Sphaerocerids generally invade a corpse at full bloat. This specimen indicated that faunal succession of this corpse occurred during a warm season with blow flies preceding the rest. The wasp nest (*Polistes* sp.) was old and of full size, indicating that the interior of the skull cranium had to be clean and dry at the time the founding wasp was searching for a nest site. The presence of a 2nd wave fly puparium (Sphaeroceridae) indicates that normal faunal succession occurred during a warm season.

In eastern Tennessee *Polistes* wasps normally seek nest sites during April. This suggested that maggot mass activity occurred prior to the late fall of the year before that. Thus, the estimated minimal postmortem interval (PMI) was set at approximately 18 months.

This conclusion was supported independently by a forensic anthropologist using skeletal weathering and by a forensic materials specialist using aging of the denim cloth.

Odontological identification of the deceased after this study was concluded showed that shc had disappeared in February, 2 years prior to discovery of the remains.

Case No. 22

Several years ago a young woman was attacked and brutally raped in a suburban Chicago wood lot by a man wearing a ski mask. Investigators quickly developed a suspect and armed with a search warrant seized a ski mask similar to the one described by the victim from the suspect's apartment. When questioned about the ski mask the suspect informed investigators that the ski mask was indeed his but that it had not been used since the previous winter.

Investigators observed large numbers of cocklebурs and other plant parts adhering to the mask. The mask and its associated vegetation was brought to a forensic entomologist for analysis. Careful dissection yielded several small caterpillars inside the cockleburs. These were subsequently identified and information concerning their development was researched.

Other studies showed that the caterpillars had a one year life cycle, and that the adult moths were active in the spring, laying eggs in early summer. Their larvae, or caterpillars, develop within the cocklebur during mid to late summer, undergo pupation during the winter, and emerge as adults the following summer. The entomologist concluded that the ski mask had been present in the outside environment during early to mid summer of that year around the time that the rape occurred. Investigators confronted the suspect armed with this information, and he subsequently confessed to having committed the rape.

In this case analysis of vegetation associated with evidence collected from a crime scene enabled investigators to accurately link the suspect to the crime scene at or about the time of the crime.

Case No. 23

In southern California on the 5th of August, the body of a 24-year old caucasian woman was found by a Search-and-Rescue Team near a dirt road in the northern rural outskirts of the Thousand Oaks Community. The nude woman had been strangled to death with her blouse, which was still tied around her neck. Several hours later a sergeant from the Ventura County Sheriff's Department conducted preliminary surveillance at the scene from approximately

10:00 p.m. to 2:00 a.m. until the coroner's investigation team arrived. The following morning the sergeant noticed a number of raised, itchy, red spots on the skin of his ankles, waist, and buttocks. Those spots were interpreted by him to be chigger bites similar to ones he obtained a number of years before while on military maneuvers in Kentucky. Later, in a discussion with members of the Search-and-Rescue Team, he learned that 20 of the 23 members also exhibited what appeared to be similar lesions. The sergeant, an outdoorsman and a long time resident of Ventura County, reasoned that these bites were unusual because chiggers were not common to southern California. Subsequently, while viewing strip-search photographs of a suspect in this case, he observed what seemed to be the same type of bites on the suspect's lower legs, buttocks, and waist. The sergeant concluded that there may be a correlation between the rarity of human-infesting chiggers in southern California, the bites on himself, the search team, the suspect, and the scene of the crime. The team met with the sergeant of the Homicide Investigation and representatives from the Ventura County District Attorney's Office. Photographs of the homicide victim, the crime site, and the suspect were reviewed. This meeting was followed by a preliminary surveillance of the crime site and surrounding areas by an entomologist. An initial visit also was made to a residence in Thousand Oaks where the suspect said he had spent the night of the homicide. During a questioning session the suspect had stated that the bites on his body were caused by fleas that he encountered at the residence of his sister who also resided in Thousand Oaks.

Investigation of the death scene area over a two day period resulted in the recovery of large numbers of unengorged, host-seeking chigger larvae immediately adjacent to the spot where the victim had been found. Evaluation of the host-parasite data indicated that the predominant concentration of host-seeking chigger larvae was in a relatively narrow zone delineated by the interface of a field disturbed by agricultural activities and the vegetation of the relatively undisturbed native plant community. A rapid decline in chigger numbers was seen at collection points outside of this zone, thus, suggesting that relatively small geographical areas provide the conditions that support the extensive chigger population levels that were observed at the homicide scene. It was found also that sampling of similar, as well as different, habitats in the surrounding area yielded few chigger larvae, and these were usually attached to a lizard host. Thus, the chance of encountering host-seeking larvae in large enough numbers to inflict multiple dermal lesions on more than 20 individuals during the same period seemed small, unless all had been at the same focus of chigger activity at about the same time.

These observations in conjunction with the testimony that the pattern and kind of dermal lesions seen in photographs of the suspect and one of the investigators were characteristic of chigger bites were presented on 31 March 1983, for the prosecution at the trial. Based on this strong circumstantial entomological evidence, the defendant was found guilty of first degree murder and was sentenced to life imprisonment without the possibility of parole.

ENTOMOLOGY IN CASES OF ABUSE AND NEGLECT

In cases involving child abuse and homicide, the forensic scientist may have little or no clues as to the circumstances surrounding the death or abuse of a child, particularly in the case of a child found in a decomposed state. The use of entomological evidence has been found to be very valuable in providing information regarding the postmortem interval, evidence of hygiene neglect and movement of the body after death.

Case No. 24

Recently, several young children were brought to a local hospital emergency room suffering from widespread diaper rash, malnutrition, and generalized neglect. A physical examination revealed that the anal and genital areas of these children were infested by maggots. Samples of the invading fly larvae were collected, preserved, and forwarded to a forensic entomologist for examination.

An analysis of the fly larvae revealed that they had been present on the children for at least 4 - 5 days. This information was used as an indicator of the minimum amount of time which had passed since the children had their diapers changed, and were properly cleaned and cared for. The entomological information was the only data available which provided quantitative information accurately determining the length of time the children had been neglected. This evidence provided crucial information in subsequent hearings concerning the welfare of the children.

Similar cases involving the analysis of insects infesting elderly patients suffering abuse and neglect at the hands of relatives and health care providers also are becoming more and more common. The potential value of entomological evidence in these matters cannot be overstated.

ENTOMOTOXICOLOGY

Recent studies have demonstrated the effects of commonly abused drugs on fly development and the importance of fly larvae as toxicological specimens. Currently research is being conducted in an effort to develop quantitative techniques for detecting and measuring drug levels in older fly puparia.

Forensic pathologists are frequently confronted with decomposed and skeletonized human remains. Oftentimes as a consequence of decomposition, traditional toxicologic specimens such as blood, urine and solid organs are no longer procurable. Recent studies have documented the value of using fly larvae as toxicologic specimens when available. So far, bromazepam, levomepromzine, malathion, phenobarbital, trazolam, oxazepam, alimemazine, clomipramine, morphine, mercury and copper have been recovered from maggots.

Case No. 25

An almost completely skeletonized corpse was discovered by walkers in a wooded area on 26 February. The decedent was a 29 year old, white male intravenous drug abuser who had been reported missing to the police on the previous 25 September, one week after he was last seen alive. His last encounter was with his girlfriend who thought he had used intravenous cocaine prior to becoming agitated. He had installed an air conditioner and then disappeared.

The body was prone, partially embedded in frozen earth in a densely wooded area. The completely skeletonized skull and mandible were located 25 feet and 8 feet from the body respectively. No drug user pharaphenalia was discovered.

At autopsy, the head and trunk were completely skeletonized. Both arms were partially fleshed and both legs and feet were completely fleshed with decomposed and desiccated skin and soft tissue. There was an irregular figure "8" hole through the pants and the right buttock from animal activity. Numerous dead, partially decomposed small and large maggots in addition to fly puparia were found on the body surface, within the clothing and in the body cavities. The clothing showed no unusual cuts, tears or holes. Complete body radiographs showed no projectiles or projectile fragments. Other than animal disruption, there was no evidence of injury.

A positive identification was made by comparison of antemortem and postmortem dental records.

Skeletal muscle from the legs along with maggots were submitted for toxicologic analysis. Both tissue and maggots tested positive for cocaine using a variety of standard toxicological procedures.

A detailed investigation into the decedent's background revealed a long history of serious cocaine addiction. Additionally, the young man had been observed staggering away from his residence in a drug induced state just prior to his disappearance some 7 months before.

Based on the totality of the investigative information and the toxicological data, the death was ruled an overdose.

Case No. 26

The shirtless body of a caucasian female of about 20 years of age was discovered at 1500 hours on 12 October in a pine woods area of NE Spokane. The victim was face-down and shoeless with several dorsal stab wounds on the left thorax. Bloodstains indicate that the victim was stabbed while lying on her right side and then rolled to the face-down position.

The corpse was in early bloat with an extensive population of maggots on face and upper ventral torso. Blackening of the face and upper body and marbling of visible extremities had occurred. The victim was lying in an open area surrounded by a scattered stand of what appeared to be ponderosa pines about 50 paces from a dirt logging road.

After 5 days refrigeration, the live maggots submitted in evidence were placed on beef kidney on a sand substrate. Adult flies began emerging on 10 November, 21 days later. Two species were present, *Cynomya cadaverina* and *Phaenicia sericata.* The former typically oviposits in the first one to two days following death. The latter ovipositеs on carrion probably within the first 24 hours and favors carrion (here a corpse) in open, sunlit sites.

Mean measurement of the preserved maggots showed three size classes. Those of medium (6 - 9 mm) are of a size that matched 7 day old larvae of these species in research done using pig carcasses under very similar weather conditions. Smaller maggots indicated the start of multiple broods of blow flies under development, and the presence of copious egg masses in the hair of the victim also supported this conclusion. One very large maggot (17.7 mm) deserves special comment. Its size indicated an age of about 3 weeks, which was not possible regarding this corpse. The remote possibility that it moved from another nearby source was ruled out because no such carrion was found by investigators, and probably more than one maggot would have been collected had such carrion been present.

An alternative explanation was that this maggot's growth was enhanced in a peculiar pocket or site in the nasal recesses of the victim. Recent research has shown very accelerated development by maggots feeding on tissues containing cocaine. A subsequent investigation revealed that the victim had a history of cocaine abuse, and a witness later testified that the decedent had snorted the drug shortly before her death.

CONCLUSIONS

Insects can be a valuable tool in investigations of homicide, suicide, untimely death, and other violent crimes. Accurate forensic determinations are possible, however, only when representative specimens are recognized, properly collected, preserved, and forwarded in a timely manner to a qualified forensic entomologist for analysis. Prosecutors, police, pathologists, and others involved in solving violent crimes should become fully aware of the complex ecology of the decay process, the important role which insects play in decomposition, and the necessity of collecting representative specimens and supplemental field data.

CHAPTER 3

ARTHROPOD BASICS - STRUCTURE AND BIOLOGY

M. Lee Goff
University of Hawaii, Honolulu, HI

E. Paul Catts
Washington, State University, Pullman, WA

The arthropods (insects and their relatives) are the earth's most widely varied organisms and account for well over half of all known species of living things. Arthropods are animals having jointed appendages (e.g., legs). They are found in virtually every area of the globe from the depths of the oceans to the highest peaks. Arthropods include the insects, crustaceans (crabs, crayfish and lobsters), spiders, centipedes, mites and other lesser known groups. Among the arthropods, insects show the greatest diversity in adaptations. This chapter is an introduction to arthropod structure and biology; in it we will briefly discuss the structure and life cycles of the groups of insects commonly associated with decomposing remains. We also will describe other arthropod groups associated with decaying corpses.

As in other arthropods, the insect has an external skeleton or exoskelton composed of chitin and protein. This exoskeleton is the body wall that gives protection for the insect's internal organs, conserves fluids and provides internal attachment sites for muscles. The exoskeleton is divided into individual sections called segments, which are variously separated by sutures or flexible joints.

The adult insect body is subdivided linearly into 3 main regions: head, thorax and abdomen (Fig. 3.1)*. The insect's head contains the brain and bears a pair of sensory antennae, a large pair of compound eyes, up to 3 simple eyes or ocelli, and the mouthparts. The antennae are quite varied in structure and range from simple filiform type of the cockroaches (Fig. 3.11) to the complex aristate type found in the muscoid flies (Fig. 3.2). The compound eyes are

* All figures are located at the end of Chapter 3.

generally quite conspicuous and are located at the upper sides of the head. The generalized insect mouthparts are of a mandibulate or chewing type, and the majority of larval insects have this type of mouthparts or some modification of them. Mouthparts in adult insects are more diverse. Adult beetles have the mandibulate type. In some of the predatory forms associated with remains, notably the Hemiptera or true bugs, the mouthparts may be modified to form an elongated beak (Fig. 3.54). Another modification of mouthparts may be found in the sponging mouthparts of adult carrion flies (Figs. 3.2, 3.39).

The second major subdivision of the insect body is the thorax (Fig. 3.1). The structures associated with locomotion (wings and legs) are all located on the thorax of the adult insect. The thorax is divided into 3 segments: from front to rear they are the prothorax, mesothorax, and metathorax (Fig. 3.1). Each of these 3 segments bears a pair of walking legs. Each of these legs will have the same general structure, but some are modified for specific functions. Common modifications of legs associated with carrion frequenting species include those needed for digging (fossorial) and grasping prey (raptorial). The mesothorax and metathorax are also the wing-bearing segments. Only adult insects will have wings, although some are wingless. The insect wings (Fig. 3.1) are outgrowths of the body wall between the top and side plates of the thorax. Wings originate as saclike structures, but become flattened when fully developed in the adult insect. A series of veins is present in the fully developed wings. These are rigid tubes and provide support and circulation for the membranous portions of the wings. In most insects, there are 2 pairs of wings originating from the mesothorax and metathorax. The Diptera, or flies, have only the front or mesothoracic pair developed as wings. Their hind "wings" are a pair of short, rodlike balancing structures called halteres (Fig. 3.2) and do not look like wings at all. There are a number of modifications of insect wings, ranging from the membranous wings of flies to the hardened front wings (elytra) of beetles (Fig. 3.13). The top portion of the prothorax often is modified to form a protective plate, the pronotum, covering the top of the mesothorax and metathorax.

The abdomen (Fig. 3.1) is the last of the 3 major subdivisions of the insect body and contains most of the insect's viscera. Externally, there is an apparent segmentation, usually with 5 to 8 segments showing. On most of these segments, there are paired lateral pores, called spiracles, that open into the internal tracheal system. A pair of spiracles also is present on the sides of the mesothorax and metathorax, respectively. Generally there will not be any obvious paired appendages on the abdomen, although a pair of terminal sensory appendages, cerci (Fig. 3.11), may be present in a few groups.

DEVELOPMENT

One of the most striking features of the phylum Arthropoda is the variety or diversity of immature forms. Growth and maturation in the Arthropoda is accomplished during a series of molts of the exoskeleton. The number and form of the various stages or growth steps within stages (instars) varies depending on the group involved. Among the insects, postembryonic development takes 1 of 3 basic metamorphic pathways: ametabolous, paurometabolous, or holometabolous.

Ametabolous metamorphosis is development "without change" and is restricted to more primitive wingless insects (the Apterygota). In this type of development, the eggs hatch into immature forms which are similar to the adults, differing only in being smaller in size and sexually immature. Among the carrion frequenting insects, the Collembola, or springtails (Fig. 3.10), and the Diplura exhibit this type of metamorphosis. Generally these insects live in the soil beneath the remains during the later stages of decomposition.

Paurometabolous metamorphosis is also termed "gradual metamorphosis." It involves a gradual change from immature to adult form. In this type of metamorphosis, both the immatures and adults occur in the same habitat and use the same food source. The egg hatches into an immature, called a nymph, which is smaller than the adult, lacks wings, but generally is similar in form to the adult. Each successive larger nymphal instar resembles the adult more in size and form. Wings develop externally as wing pads which increase gradually in size and then expand during the final molt to the adult stage. Among the carrion-frequenting insects, cockroaches (Fig. 3.11 - Blattaria) and the various predatory bugs (Fig. 3.54 - Hemiptera) exhibit this type of metamorphosis.

The most complex type of metamorphosis is holometabolous or complete metamorphosis. Generally an egg is laid (oviposited), although a few, such as the flesh flies (family Sarcophagidae), deposit larvae (larviposition) rather than eggs. The larvae, either hatching from the egg or deposited by the adult female, increase in size by growth steps (instars) and moults, finally entering an apparently inactive pupal stage. During the pupal stage, the adult insect develops internally and eventually emerges from the pupal case. The larval stage in holometabolous insects is quite different in appearance from the adult and is adapted to use different habitats and food sources than the adult. Several different forms of larvae can be distinguished easily in the holometabolous insects (Figs. 3.23, 3.34, 3.41). The larval stage eventually develops into the pupa (plural is pupae). The pupal stage gives rise to the adult insect. There are 3 types of pupae recognized in insects exhibiting holometabolous metamorphosis. In the exarate type of pupa, the appendages (legs, antennae, and wings) of the developing insect are free from the body and can be recognized easily. This type of pupa is generally pale in color and resembles a mummified adult. This is the most common type of pupa in insects, but is not found in the flies (Diptera) and butterflies (Lepidoptera). In the obtect type of pupa (Figs. 3.33, 3.35), the appendages are glued to the body. This type of pupa is encountered in the primitive or lower flies (Diptera) and in the moths and butterflies (Lepidoptera). In the Lepidoptera, the pupa may be enclosed in a protective silken cocoon produced by the larva prior to pupation. The coarctate type of pupa (Fig. 3.49) is similar to the exarate type, but is enclosed inside of the last larval skin, hardened to form a protective shell called a puparium. This is the pupal type of many of the more advanced or higher flies (Dipera) associated with carrion.

The length of time required for development varies considerably with the kind of insect. Many kinds of insects are capable of having several to many generations within a single year. Environmental factors such as temperature, day length, available food supply, and humidity may serve to limit the times of the year during which insects can develop. Thus, the species and forms of insects present on or under the corpse will vary with the geographic location and the time of year.

ARTHROPODS FREQUENTLY ASSOCIATED WITH CARRION

A number of different arthropods, including insects, may be recovered during the examination of remains. Not all of these groups (taxa) will be of equal significance in estimating the postmortem interval. However, collectively, they may yield data concerning the events related to the death and decay of the remains. Some taxa will be incidental, having simply wandered onto the remains from the surrounding area (e.g., aphids). Some groups, such as ants, spiders, and centipedes (Figs. 3.4, 3.8, 3.52), may be predators, feeding on species exploiting the remains and thus having an impact on the size of other groups there. Other taxa, such as fly maggots, feed on the remains proper and will be associated intimately with the decomposition process. This latter group is of major significance in estimating the postmortem interval. All arthropod taxa in all stages available should be identified, collected and preserved in order to present as complete a picture of the condition of the remains as possible. Only brief descriptions of the major taxa commonly encountered in association with remains is presented here. Keep in mind that this is only an overview and that situations and groups or species represented will vary considerably with the geographic area involved and with the location and condition of the corpse.

ABOUT THE ILLUSTRATIONS

The illustrations of insects in this book are intended to show the general appearance of those arthropods more often associated with a decomposing corpse. These are all creatures which we have found with dead bodies. Although actual specimens were used as models, the illustrations are not meant to be accurate to the last wing vein, bristle and gnat's eyelash. However they will serve a useful purpose for both the death scene investigator and the entomologist as an aid to recognition of the diversity of small animals that find their life after our death.

In all illustrations of insects and other arthropods, the head end of the creature is either at the top or the left of the drawing. Also, the approximate overall length is indicated by the letter "L." For immature stages the size given is for the fully grown final instar.

ARTHROPODS (phylum Arthropoda):

CHELICERATES (subphylum Chelicerata):

ARACHNIDS (class Arachnida - 10 orders):

MITES (order Acari): Figs. 3.3, 6.4)

Characteristics: Small organisms (generally less than 1 mm in length); body divided into 2 regions (small head and larger body with legs); adults and nymphs with 4 pairs of legs; larvae with only 3 pairs of legs; no wings.

Mites are a significant component of the arthropod assemblage in the soil under remains during the later stages of decomposition. Most mites are small and inconspicuous. Some, such as the Macrochelidae, attach to adult flies for transport. These species are carried to the remains early in the decomposition process and feed on the eggs and small maggots and the larvae of other insects. Other kinds of mites are soil dwelling forms, living under the remains where they feed on other small soil arthropods and their eggs associated with the remains. These include many of the predatory forms, such as Cunaxidae (Fig. 3.3). Another group of mites consists of fungus feeders or detritus feeders. These include the Acaridae and Winterschmidtiidae. They feed in the soil under the remains and large populations are common during the later stages of decay. Due to their small size, mites are generally not observed on the remains and specialized collecting techniques should be employed. Generally, most of the species will be found in soil samples from the seepage area below the remains.

SPIDERS (order Araneae): (Fig. 3.4)

Characteristics: Size variable (length generally less than 75 mm); body divided into cephalothorax (combined head and leg-bearing regions) and unsegmented abdomen; 3 or 4 pairs of simple eyes on cephalothorax; mouthparts modified to form fangs.

Spiders are associated with decomposing remains as predators of other arthropods. All of the species reported to occur also are commonly recovered from other situations. No instance of feeding by spiders on the carrion or decomposition fluids has been documented.

PSEUDOSCORPIONS (order Pseudoscorpionida): (Fig. 3.5)

Characteristics: Small in size (length generally less than 5 mm); body divided into cephalothorax and segmented abdomen; pedipalps and mouthparts chelate (i.e., like pinchers); 4 pairs of legs.

These small organisms are common soil-dwelling predators of other arthropods. Their association with remains is related to their predatory habits only. These organisms may be common during the later stages of decomposition (Early and Goff 1986).

SCORPIONS (order Scorpionida): (Fig. 3.6)

Characteristics: size variable, up to 125 mm in length; abdomen segmented, and terminating in a sting; pedipalps long and chelate (i.e., like pinchers); 4 pairs of legs.

Scorpions are primarily nocturnal. Their association with remains is as opportunistic predators of other organisms found there. Payne and Crossley (1966) have suggested that they may feed on decomposing tissues or other fluids.

CRUSTACEANS (subphylum Crustacea):
(class Malacostraca):
PILL BUGS (order Isopoda, Fig., 3.7)

Characteristics: Body somewhat flattened; 7 pairs of thoracic appendages; 2 pairs of antennae; chewing (mandibulate) mouthparts.

Pill bugs are frequently associated with decomposing remains and may be recovered during all stages of decomposition (Early and Goff 1986; Tullis and Goff 1987). Usually they are associated with rotting vegetation although they are known to feed on decomposing animal tissues (Payne and Crossley 1966).

MANDIBULATES (subphylum Atelocerata):
CENTIPEDES (class Chilopoda, Fig. 3.8)

Characteristics: Size variable (some temperate species up to 150 mm in length; tropical species up to 0.5 m in length); body elongate, wormlike, and segmented; each segment bears one pair of legs; antennae long with 14 or more segments; chewing mouthparts (mandibulate); appendages of 1st postoral segment modified to form poison jaws.

Centipedes are frequently recovered from remains where they are predatory on other small animals associated with the remains.

MILLIPEDES (class Diplopoda, Fig. 3.9)

Characteristics: Body elongate, wormlike, segmented; most segments bear 2 pairs of legs; antennae short (with less than 10 segments.)

The millipedes are generally found in moist protected habitats, and most species are believed to be plant feeders. Observations by

Hoffman and Payne (1969) indicate that a number of species feed on animal tissues, including decomposing remains.

INSECTS (class Insecta):

Characteristics: Body divided into 3 major sections; 3 pairs of walking legs; 1 pair of antennae; wings present in adults or absent.

Among the arthropods associated with decomposing remains, the insects are by far the predominant group. They are the first group to arrive at the remains and will be among the last to depart. In most of the studies conducted to date, the insects have accounted for over 85% of the arthropods encountered. Due to the importance of this group in forensic studies, accounts will be presented of the various orders commonly encountered.

WINGLESS INSECTS (subclass Apterygota):
SPRINGTAILS (order Collembola, Fig. 3.10)

Characteristics: Small insects, less than 7 mm in length; chewing mouthparts; eyes consist of lateral patches of simple eyes (ocelli); antennae short (generally 4-segmented); wings absent; jumping structures (a lever and trigger) located on abdomen.

Springtails are small insects which get their name from their ability to jump using a combination of structures, the furcula and retenaculum, arising terminally and ventrally from the abdomen, respectively. These insects are common soil inhabitants throughout most of the year, even in cold climates. They appear to be primarily plant feeders, although some species do feed on decaying animal material. Habitats for these insects are principally damp and, in this respect, a corpse and underlying seepage provides an ideal temporary habitat. Species associated with decaying remains do not appear to be specific to this habitat, but represent species normally encountered in surrounding soil and leaf litters.

DIPLURANS (order Diplura):

Characteristics: Minute to small insects (less than 7 mm in length); chewing mouthparts; eyes absent; antennae long, with more than 10 segments (filiform); wings absent; abdomen terminating in a pair of appendages (cerci).

These small arthropods are most frequently associated with rotting wood or other vegetable material. None of the species appear to be associated specifically with remains but do occur in the soil under the remains.

SILVERFISH (order Thysanura):

Characteristics: Small insects (9-14 mm in length) having the body covered with scales; chewing mouthparts; compound eyes usually present; antennae long, filiform; wings absent; abdomen terminating in 3 appendages (a pair of cerci and a median caudal filament).

Silverfish are common pests of homes and stored products. Association with decaying remains appears to be opportunistic and most frequent during the later stages of decomposition when the remains have dried (Early and Goff 1986).

WINGED INSECTS (subclass Pterygota):
COCKROACHES (order Blattaria, Fig. 3.11)

Characteristics: Body shape an elongated oval, flattened; head concealed beneath pronotum; chewing mouthparts; compound eyes well developed; antennae long, filiform; 2 pairs of wings, the first pair leathery (tegmina), second pair concealed under the first; with gradual metamorphosis.

Cockroaches are widely distributed, omnivorous insects. Their association with remains appears to be opportunistic. They feed on decomposing tissues and fluids. Although they are not associated with any particular period in the decomposition process, they are uncommon at fresh carrion (Early and Goff 1986). Activity by cockroaches at carrion is nocturnal (Payne and Crossley 1966).

BEETLES (order Coleoptera):

Characteristics: Size variable (from minute to large); chewing mouthparts; compound eyes present or absent, ocelli generally absent; antennae of variable form; 2 pairs of wings, front pair hardened as protective shields (elytra), hind pair membranous and usually folded under the elytra; larvae of those found in carrion are wormlike with 3 pairs of thoracic legs; with complete metamorphosis.

The Coleoptera is the largest order (40%) of all insects in number of species. It contains many groups of significance to forensic study. Due to the different roles played by various species of beetles, there is no characteristic time of appearance for this order in the decomposing process. The predatory forms, such as the Staphylinidae (Figs. 3.14, 3.23), may arrive within a few hours after death and remain active well into the later stages of decomposition (Early and Goff 1986). In South Carolina, Payne (1965) noted scarab beetle adults (Figs. 3.17, 3.25) arriving at remains on the second day of exposure and similar observations were made by Reed (1958) during his studies in Tennessee. Catts (pers. comm.) collected them on a pig within 15 minutes following death. These scarabs are primarily scavenger species frequently associated with dung and decaying plant materials. Carabid, or ground beetles, (Fig. 3.21) are predatory both as larvae and adults and may be found with remains during all stages of decomposition. Another predatory group is the family Histeridae (hister beetles) (Figs. 3.20, 3.28). These beetles arrive early in the decomposition process and feed primarily on maggots (Payne and Crossley 1966).

The Silphidae (carrion beetles or burying beetles) (Figs. 3.13, 3.16, 3.24) are a conspicuous group in many areas, arriving at remains during early decay (Payne 1965) and remaining as adults and immatures until the dry stage. There have been conflicting accounts of the food habits of silphids. Reed (1958) stated that the immatures fed on the remains and occasionally on maggots, while adults fed only on maggots, with some exceptions. Payne and Crossley (1966) found adult silphids feeding only on maggots and made no mention of immatures feeding. Regardless of feeding habits, these beetles are intimately associated with the decomposition process.

Characteristic of the beginning of the dry stages of decomposition at many sites is the arrival of the skin beetles (Dermestidae) (Figs. 3.19, 3.29). These beetles feed primarily on dried skin and tissues both as adults and larvae. In addition, the dermestids are fed upon by species of checkered beetles (Cleridae) (Figs. 3.18, 3.27). Representatives of the Tenebrionidae (darkling beetles, Figs. 3.22, 3.26) also are recorded from remains, but appear to feed on materials other than the remains proper (Payne and Crossley 1966).

There are a number of additional families of Coleoptera reported from decomposing remains. Payne and Crossley (1966) listed 40 families from the southeastern United States, and Goff et al. (1986)

listed 17 from the Hawaiian Islands. While each of these families may be of some significance for a given case, the groups identified above are common to most death scenes.

EARWIGS (order Dermaptera, Fig. 3.12)

Characteristics: Small to medium insects (less than 2 cm in length); body elongate; chewing mouthparts; compound eyes generally present; antennae long with many segments, filiform; 2 pairs of wings, front pair short and thickened (tegmina), hind pair membranous, folded under front wings; abdomen with a pair of unsegmented, terminal, pincherlike structures (modified cerci).

The earwigs feed primarily on decaying plant and animal materials although a few species are predatory. While not specific to decomposing remains, they may occur in large numbers on and under a corpse, particularly in wet conditions (Tullis and Goff 1987).

TRUE FLIES (order Diptera):

Characteristics: Adult with the anterior pair of wings membranous and well developed, but the posterior wings modified to form rodlike stumps (halteres); mouthparts variously modified for sucking or sponging fluids; antennae vary from filiform to aristate; compound eyes present; larva wormlike or maggotlike, without legs or other appendages; pupa coarctate (higher flies) or obtect (lower flies).

The order Diptera is the fourth largest order of insects in number of species. Flies occupy a wide variety of habitats and are the predominant group of insects associated with early stages of decomposition. The order is divided into 2 suborders: Nematocera (lower flies) and Brachycera (higher flies). While there are families in both suborders associated with remains, those of greater significance to forensic sciences are among the higher flies.

Generally the Nematocera are considered to include primitive flies. Adults of this suborder have antennae which are filiform and longer than the head. Larvae (Figs. 3.32, 3.34, 3.36) usually occur in aquatic or moist habitats. Larvae have well developed heads, opposable mandibles, and a discrete thorax. Nematocera pupae are obtect and frequently motile. Families of Nematocera found with remains include Tipulidae (crane flies), Trichoceridae (winter crane flies, Fig. 3.30), Psychodidae (moth flies), Ceratopogonidae (biting midges), Chironomidae (midges, Figs. 3.34, 3.35), Mycetophilidae (fungus gnats), Sciaridae (dark-winged fungus gnats), and Scatopsidae (scavenger flies). Most of these have been reported only as adults associated with remains, although Tullis and Goff (1987) and Payne et al. (1968) reported Psychodidae larvae as a significant component of the decomposition process in warm, wet tropical habitat.

Adult Brachycera are characterized as small to large flies having antennae with fewer than 5 segments (generally 3-segmented). Adults of many species are predators or parasites of other insects, thus accounting for their attraction to remains. The larvae (Figs. 3.31, 3.41) generally have the head either not sclerotized, or incompletely sclerotized and retractile into the body. The pupa (Figs. 3.31, 3.49, 3.50, 3.51) is coarctate, and the adult fly emerges through a "T"-shaped opening or through a circular cap at the head end of the puparium. Families of more primitive Brachycera having species reported from remains include Stratiomyidae (soldier flies, Fig. 3.31), Scenopinidae (window flies), Asilidae (robber flies), and Dolichopodidae (long-legged flies). Of these families, only the Stratiomyidae and Scenopinidae appear to be of significance. Stratiomyidae are present as adults feeding on decomposition fluids during the early part of the decay process and as larvae during the middle and later stages of decomposition. Scenopinidae appear during the later stages of decomposition preying on soil dwelling arthropods under the remains.

The more advanced higher flies may be characterized as relatively stout-bodied flies with 3-segmented, aristate antennae (Fig. 3.2). The larvae are maggotlike, lacking a sclerotized head. In the larva, the primary respiratory openings (spiracles) (Fig. 3.41, 3.43, 3.45) are located in a pair of sclerotized spiracular plates located at the tail end of the maggot. These spiracular plates are often distinctive for family or generic level identifications. During the earlier stages of decomposition, larvae (maggots) and adults of these flies are the predominant arthropods associated with the remains. Most of the species of major significance are found in 3 families (Figs. 3.2, 3.41, 3.42, 3.43, 3.44): Calliphoridae (blow flies), Muscidae (house flies), and Sarcophagidae (flesh flies). Scarcophagid flies were the first species to arrive at carrion in studies conducted in South Carolina and Hawaii (Payne 1965; Early and Goff 1986; Tullis and Goff 1987), followed by species of Calliphoridae. But in cooler, temperate regions, the Calliphoridae arrive first (Reed 1958; Rodriguez and Bass 1983; Smith 1986). Generally, the blow flies deposit eggs and the flesh flies larvae in areas around exposed natural body openings (eyes, nostrils, ears, mouth, and less at anus, and genitals) or around wounds. The Muscidae follow a similar pattern but are later to arrive. Although some species of Sarcophagide and Calliphoridae are known to be predatory or parasitic, the predominant role for carrion-fequenting species is scavenging. Maggots of blow flies and flesh flies look very similar. However, they can be separated based on their posterior spiracles. Among blow flies the inner slit of each spiracle slants inward, and in flesh flies the inner slit slants outward (see Figs. 3.41 and 3.42). This condition holds true for both the 2nd (L II) and 3rd (L III) instar larvae.

There are a number of additional families of flies reported from remains. Most are groups associated with rotting materials and their significance to determination of postmortem intervals will vary with individual cases. In damp habitats, Phoridae (hump-backed flies, Fig. 3.48), Piophilidae (skipper flies, Figs. 3.39, 3.47), Sepsidae (black scavenger flies, Fig. 3.45), and Sphaeroceridae (small dung flies, Figs. 3.40, 3.46) may be conspicuous components of the fauna associated with remains (Reed 1958, Tullis and Goff 1987). Species in families not generally associated with decomposing animal materials also may be attracted to remains in the adult stage. Adult Drosophilidae (fruit flies), normally associated with ripe fruit and decaying vegetation were observed on decomposing pig carcasses (Payne and Crossley 1966; Tullis and Goff 1987). Species of Syrphidae, normally associated with flowers, also have been observed feeding on decomposition fluids (Payne and Crossley 1966).

TRUE BUGS (order Hemiptera, Fig. 3.54)

Characteristics: Body minute to large; anterior pair of wings half membranous, half thickened (hemelytra), posterior wings membranous; mouthparts as a piercing-sucking beak, arising from front of head and extending back along the venter; antennae usually long, 4-5 segments; compound eyes generally well developed; with gradual metamorphosis.

Species of Hemiptera associated with remains are in families generally associated with plant materials or predatory forms. Plant feeders reported from carrion include Coreidae, Cydnidae, Lygaeidae, and Miridae (Payne and Crossley 1966; Early and Goff 1986; Tullis and Goff 1987). Two families of predatory Hemiptera are reported commonly from carrion, both feeding on maggots: Reduviidae (Fig. 3.54) and Nabidae (Payne and Crossley 1966; Tullis and Goff 1987).

ANTS, BEES, AND WASPS (order Hymenoptera, Figs. 3.52, 3.53)

Characteristics: Adult body minute to large; most with 2 pairs of membranous wings, absent in some forms; mouthparts variable, chewing, lapping or sucking forms; antennae filiform or geniculate (elbowed) (Fig. 3.52); compound eyes well developed, 3 dorsal ocelli usually present; first abdominal segment fused with metathorax, constricted posteriorly to form a waist between thorax and abdomen; females with ovipositor frequently modified to pierce plant tissues or as a sting; complete metamorphosis; larva wormlike or caterpillar-like; pupa exarate, usually within a cocoon.

A number of different families of Hymenoptera have been reported from remains. Payne and Mason (1971) listed 82 species associated with pig carrion in South Carolina. Relationships of these species to carrion differ, ranging from direct feeding on the corpse, through predation, to parasitism of other insects exploiting remains. Ants as predators can be a significant factor in the decomposition process and may retard decomposition by decreasing the maggot population (Early and Goff 1986). Bees and wasps have been observed feeding directly on the fluids and tissues during the earlier decomposition stages. Wasps, especially the Vespidae, (Fig. 3.53) are frequently predatory on fly adults and larvae present on remains (Payne 1965; Early and Goff 1986). There are several families of Hymenoptera observed at remains which are parasites of larvae and pupae of Diptera, Coleoptera, and other insects. These families include the Bethylidae, Brachonidae, Chalcidae, Diapriidae, Encyrtidae, Evaniidae, Ichneumonidae, Pelecinidae, Proctotrupidae, and Sclerogibbidae. This is by no means a complete listing of the parasitic Hymenoptera associated with remains but serves only to emphasize the diversity of this assemblage associated with decomposing carrion. They have little significance on the postmortem interval estimate, except where, in large numbers, through predation they may indirectly affect the activity of other insects, especially maggots.

BUTTERFLIES AND MOTHS (order Lepidoptera):

Characteristics: Adult small to very large, covered with scales; 2 pairs of scaled membranous wings; antennae prominent, variable in form; mouthparts generally a coiled sucking proboscis (vestigial in some species); compound eyes prominent; complete metamorphosis; larva caterpillar-like, abdomen usually with 5 pairs of prolegs; pupa obtect, may be enclosed in cocoon.

Adult Lepidoptera are frequently attracted to decaying remains and feed on decomposition fluids. Reed (1958) listed 14 species of adult Lepidoptera associated with dog carcasses in Tennessee, 10 of which he observed feeding on the carcasses. These were in the families of Geometridae, Hesperidae, Lycaenidae, Nymphalidae, Papilionidae, Phalaenidae, and Sphingidae. Similar results were given by Payne and King (1969). Early and Goff (1986) and Tullis and Goff (1987), working in both wet and dry habitats in the Hawaiian Islands, recovered larvae of Tineidae and Gelechiidae from soil under carcasses during later decomposition.

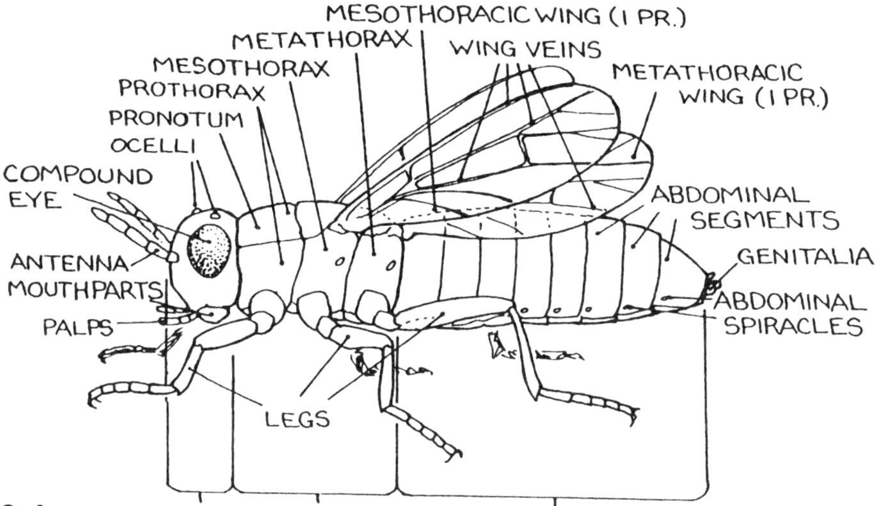

3.1

PARTS OF A HYPOTHETICAL ADULT INSECT

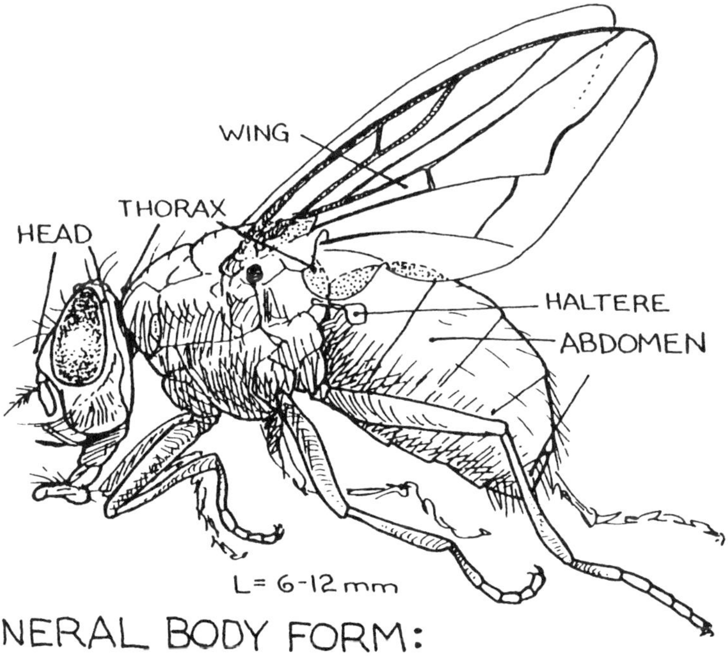

3.2

GENERAL BODY FORM:
BLOW FLY, FLESH FLY AND HOUSE FLY

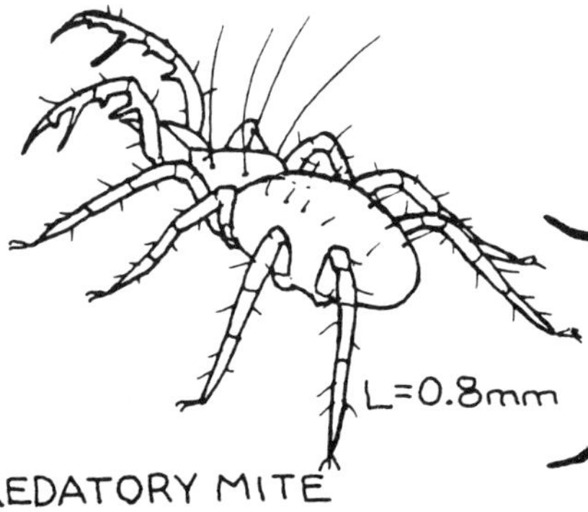

PREDATORY MITE
ACARINA-CUNAXIDAE

3.3

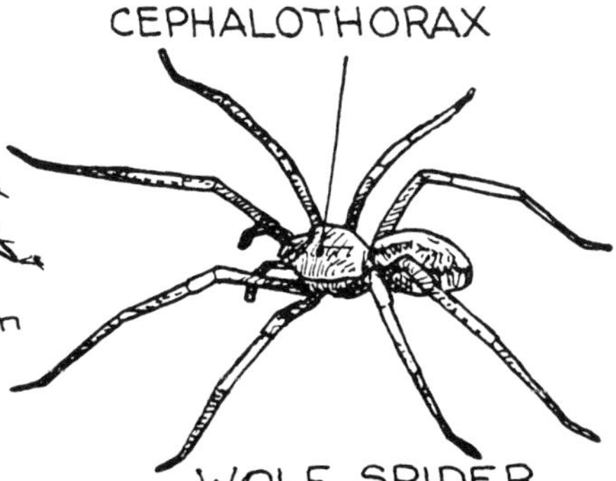

3.4 WOLF SPIDER
ARANEAE
LYCOSIDAE
L=<75mm

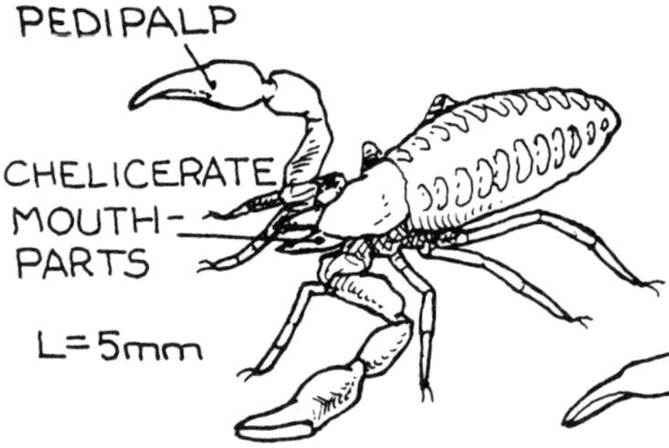

PSEUDOSCORPION
PSEUDOSCORPIONIDA

3.5

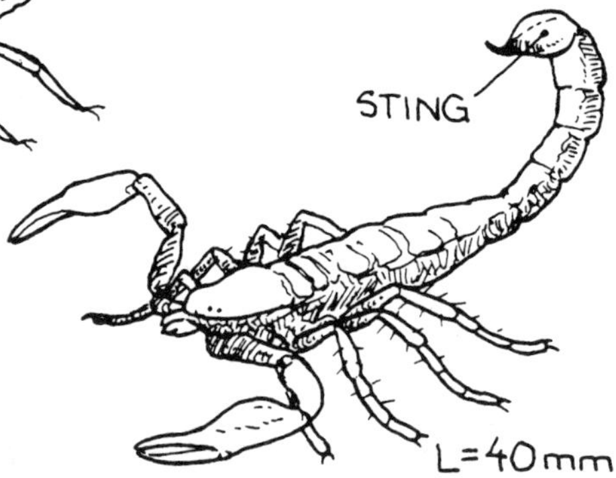

3.6 SCORPION
SCORPIONIDA

SOME NONINSECT PREDATORS
ASSOCIATED WITH CARRION

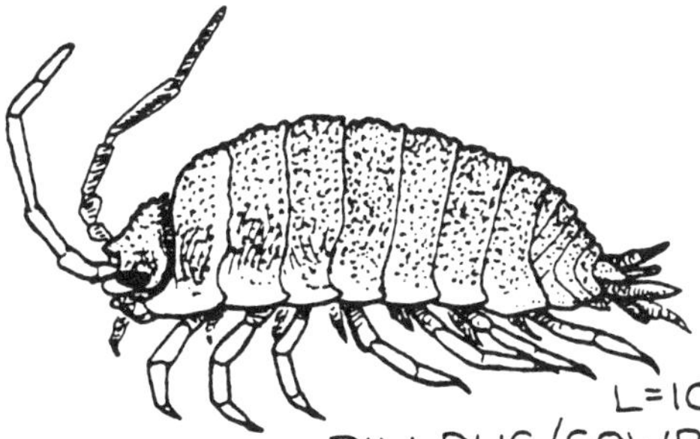

L=10mm

3.7 PILLBUG/SOWBUG ISOPODA

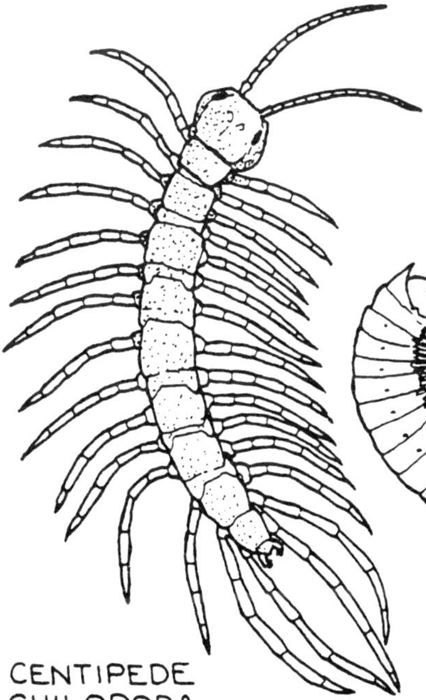

CENTIPEDE CHILOPODA L=30-40 mm

3.8

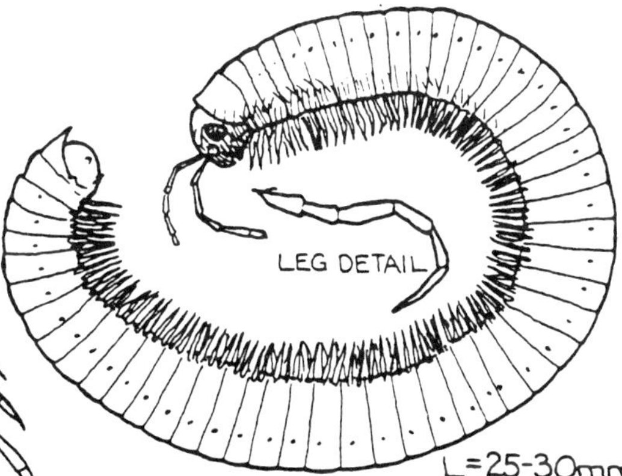

L=25-30mm

MILLIPEDE - DIPLOPODA

3.9

WORMLIKE ARTHROPODS ASSOCIATED WITH CARRION

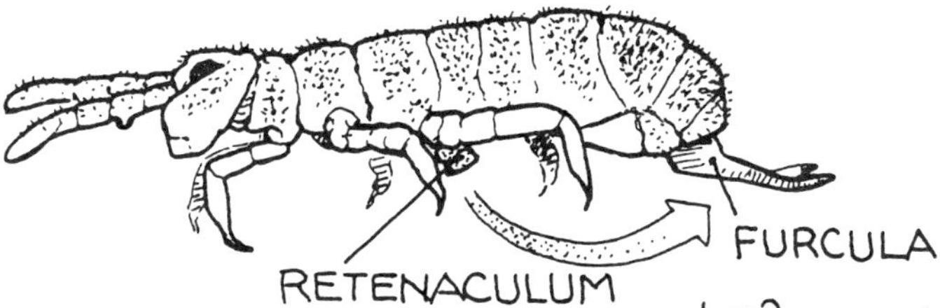

L=2mm

SPRINGTAIL - COLLEMBOLA
ISOTOMIDAE (Proisotoma sp.)
3.10

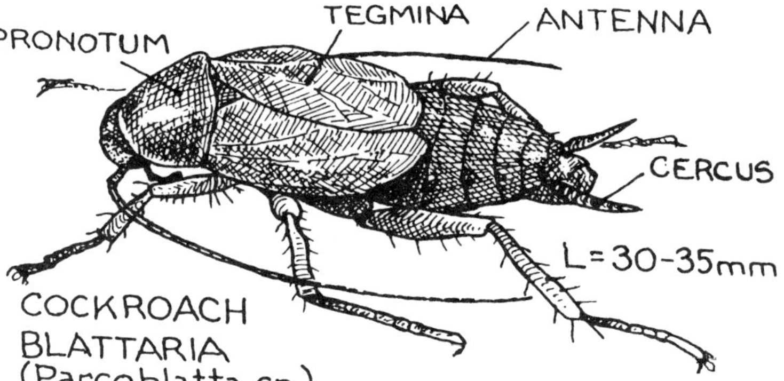

COCKROACH
BLATTARIA
(Parcoblatta sp.)
3.11

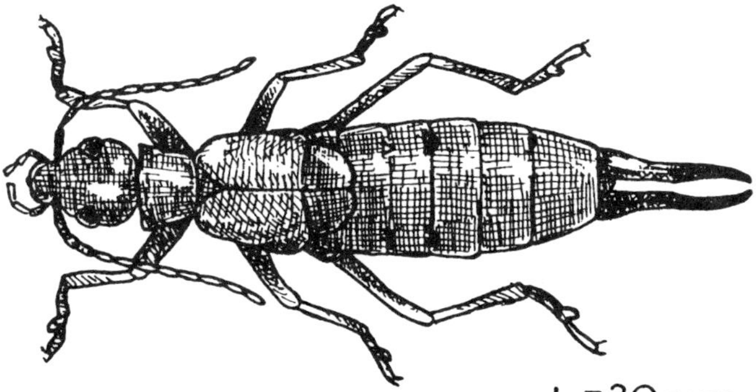

L=20mm

EARWIG - DERMAPTERA
3.12

FAST RUNNING AND JUMPING INSECTS ASSOCIATED WITH CARRION

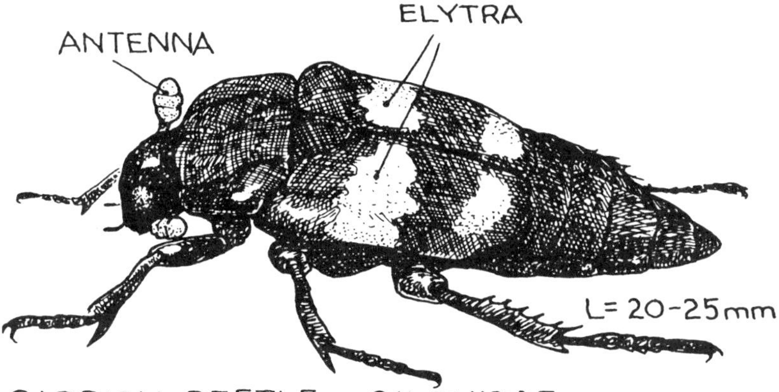

CARRION BEETLE - SILPHIDAE
(*Nicrophorus* sp.)

3.13

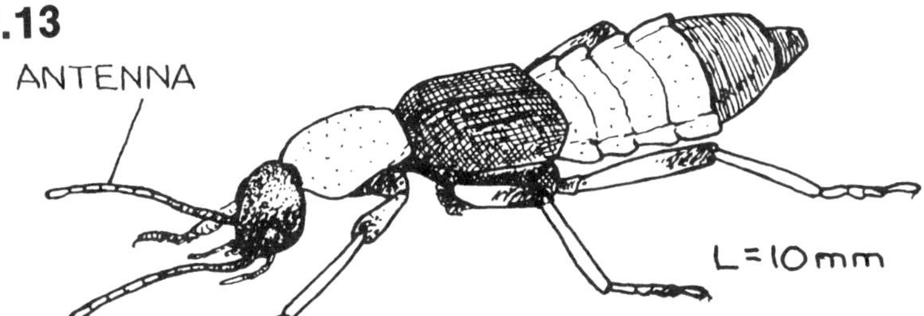

ROVE BEETLE
STAPHYLINIDAE (*Lathrobium* sp.)

3.14

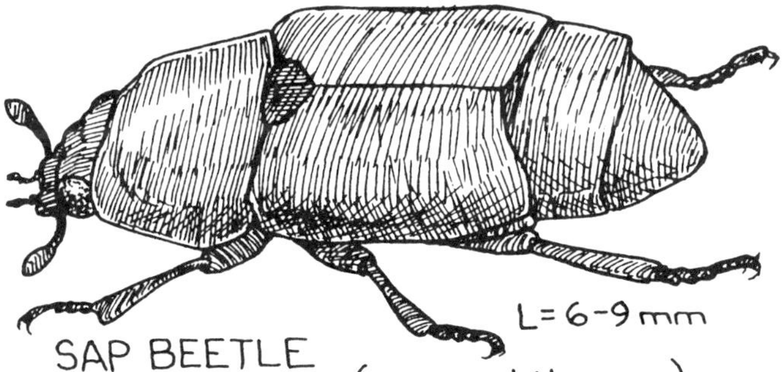

SAP BEETLE
NITIDULIDAE (*Carpophilus* sp.)

3.15

LARGER BEETLES ASSOCIATED WITH CARRION

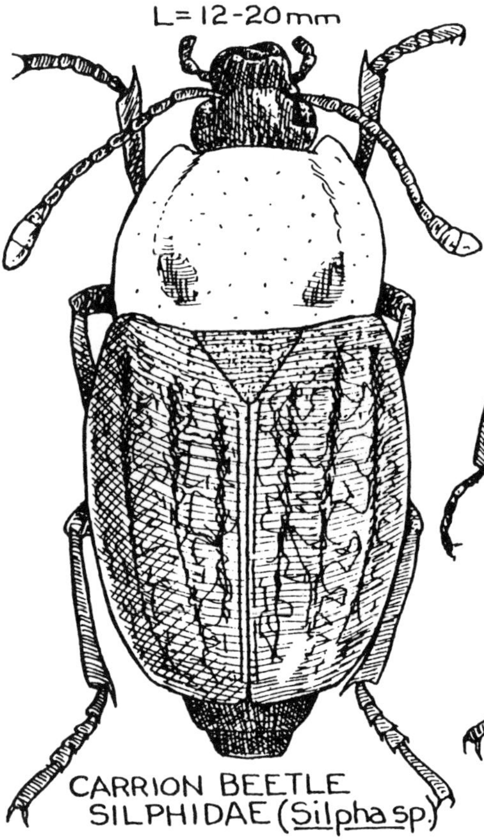

CARRION BEETLE
SILPHIDAE (*Silpha* sp.)

3.16

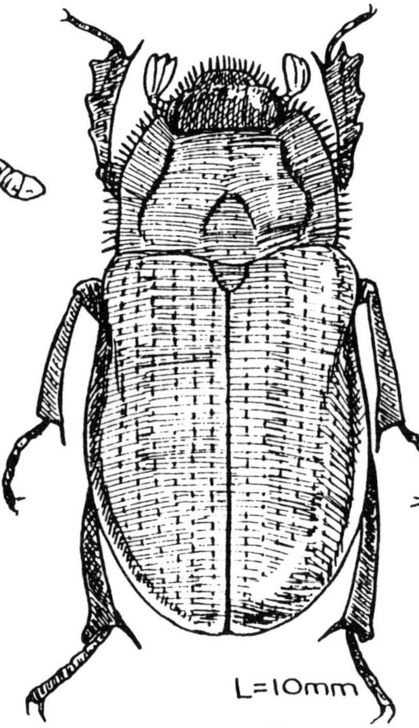

HIDE BEETLE
TROGINAE (*Trox* sp.)
SCARABAEIDAE

3.17

MORE LARGE BEETLES ASSOCIATED WITH CARRION

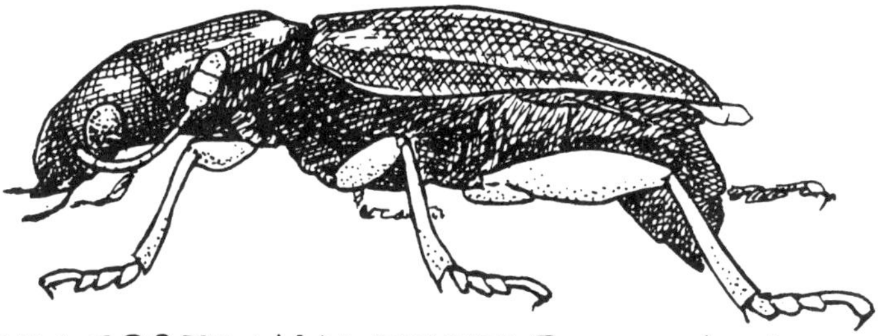

RED-LEGGED HAM BEETLE L=6mm
CLERIDAE (*Necrobia rufipes*)
3.18

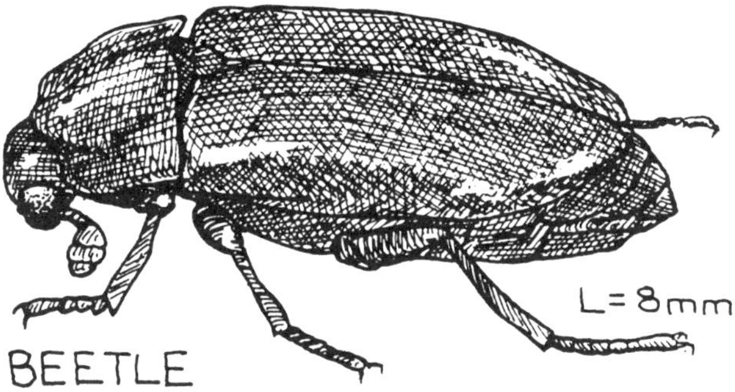

SKIN BEETLE
DERMESTIDAE
3.19

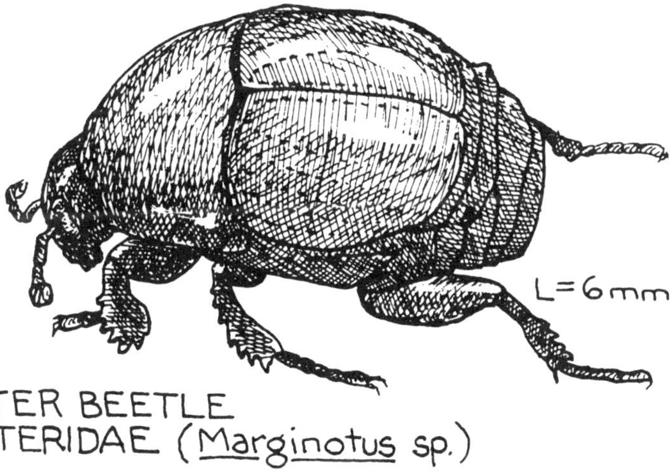

HISTER BEETLE
HISTERIDAE (*Marginotus* sp.)
3.20

SMALLER BEETLES
ASSOCIATED WITH CARRION

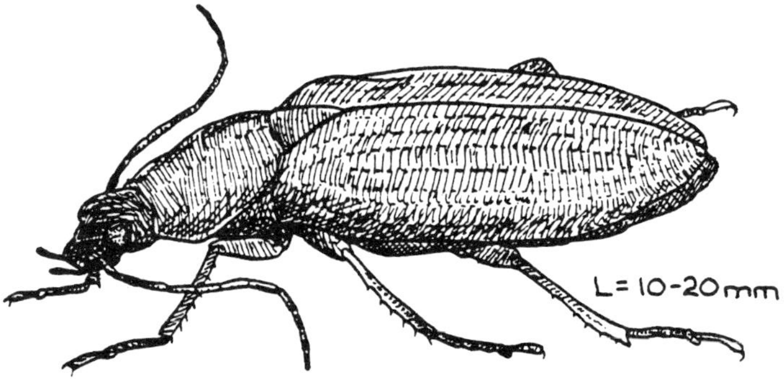

GROUND BEETLE – CARABIDAE
3.21

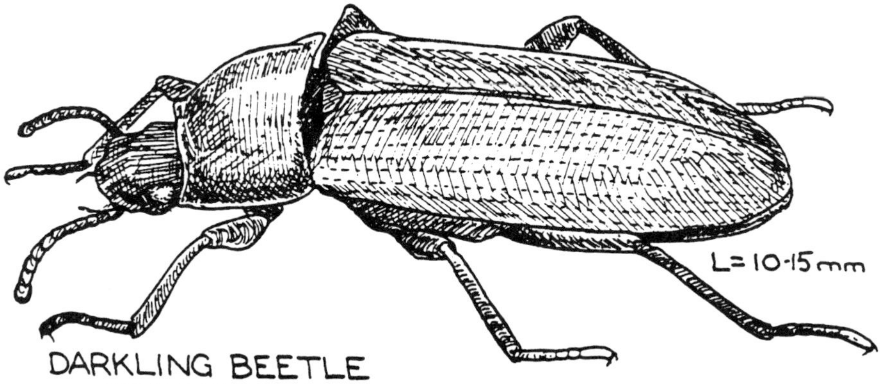

DARKLING BEETLE
TENEBRIONIDAE
3.22

OTHER BEETLES ASSOCIATED WITH CARRION

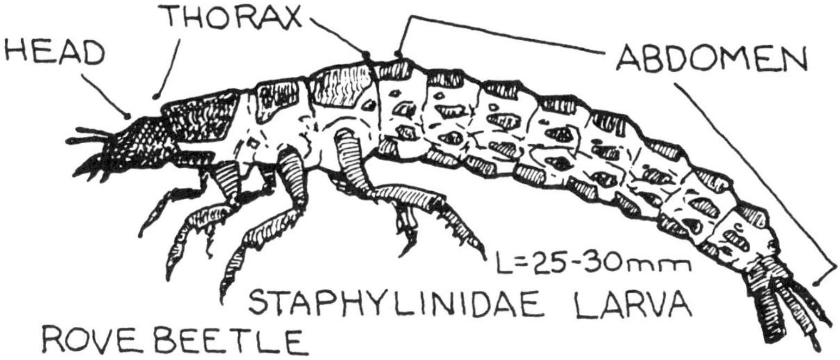

STAPHYLINIDAE LARVA
ROVE BEETLE
3.23

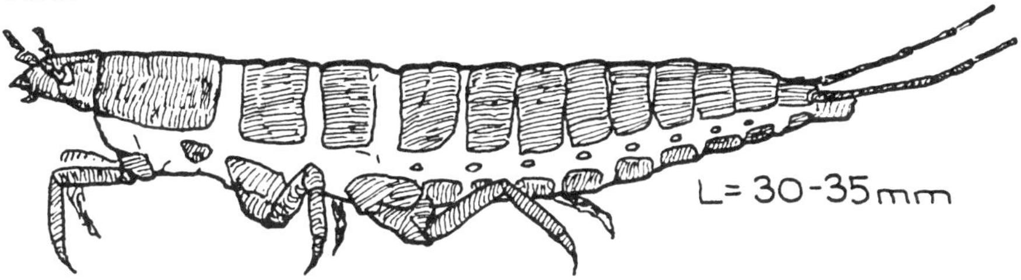

SILPHIDAE LARVA
CARRION BEETLE
3.24

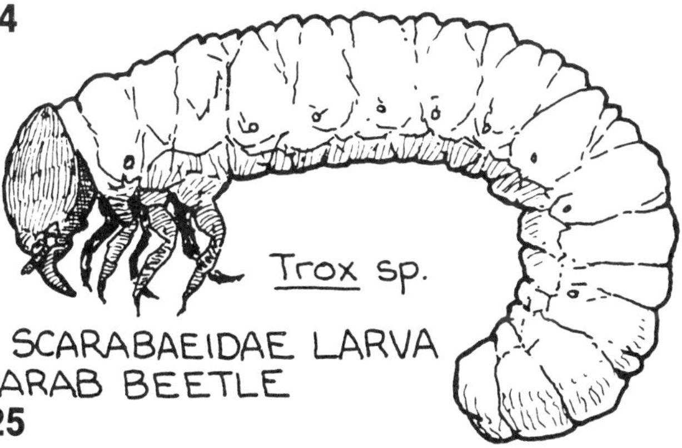

SCARABAEIDAE LARVA
SCARAB BEETLE
3.25

TENEBRIONIDAE LARVA
DARKLING BEETLE
3.26

LARGER BEETLE LARVAE ASSOCIATED WITH CARRION

3.27

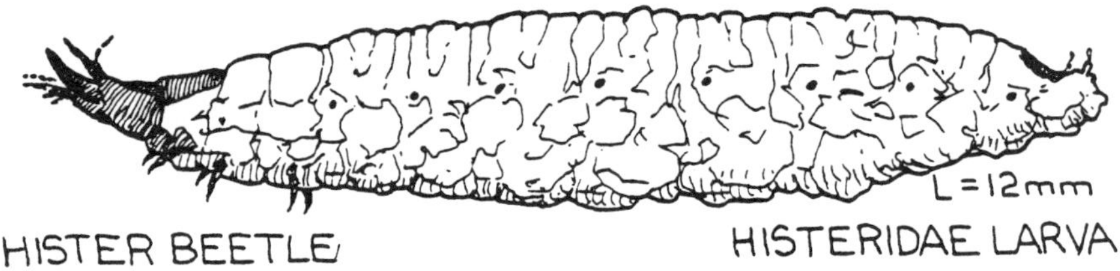

3.28

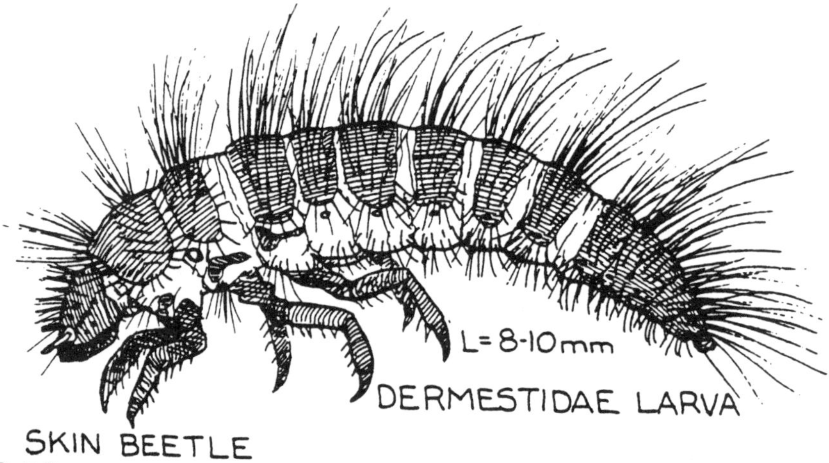

3.29

SMALLER BEETLE LARVAE
ASSOCIATED WITH CARRION

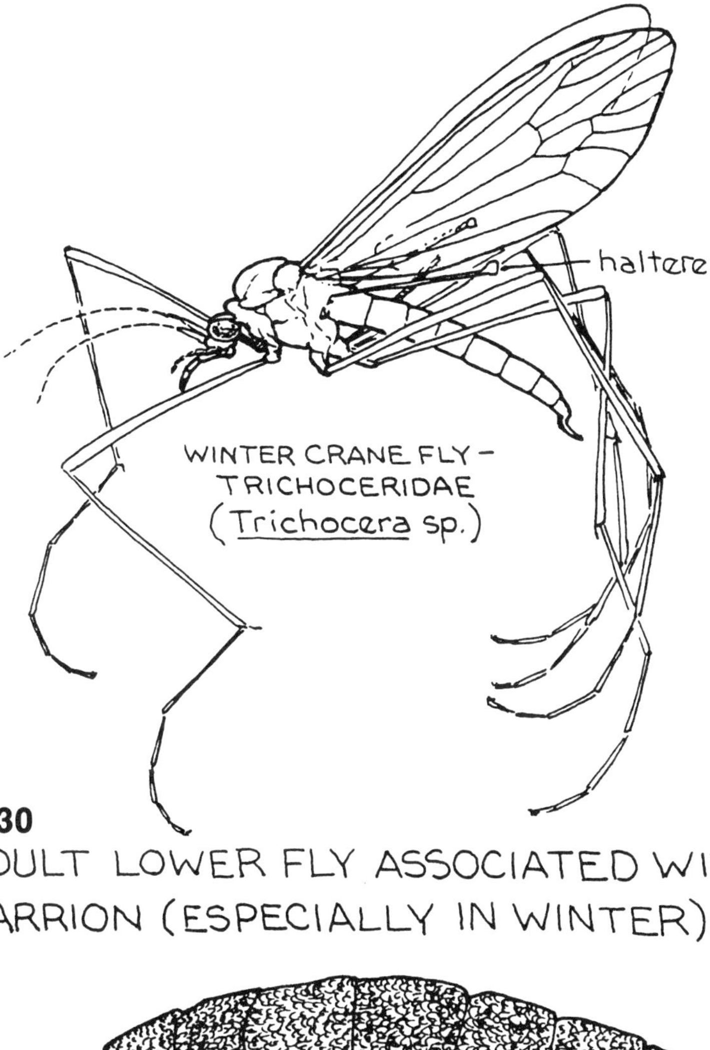

3.30
ADULT LOWER FLY ASSOCIATED WITH CARRION (ESPECIALLY IN WINTER)

STRATIOMYIDAE LARVA L=18-23mm
(PUPA IS SIMILAR) *Hermetia illucens*

3.31
BLACK SOLDIER FLY MAGGOT

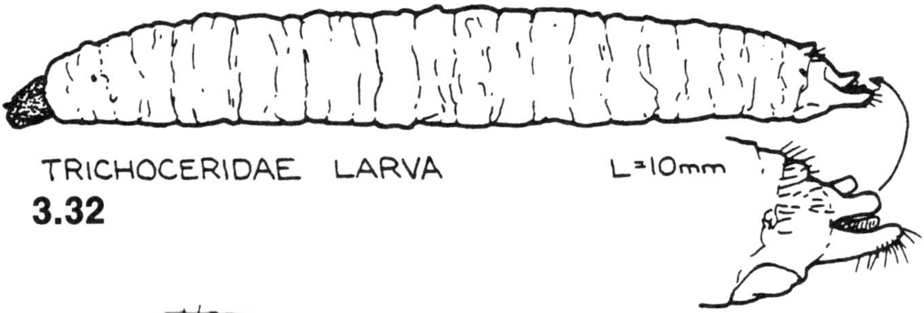

TRICHOCERIDAE LARVA L=10mm

3.32

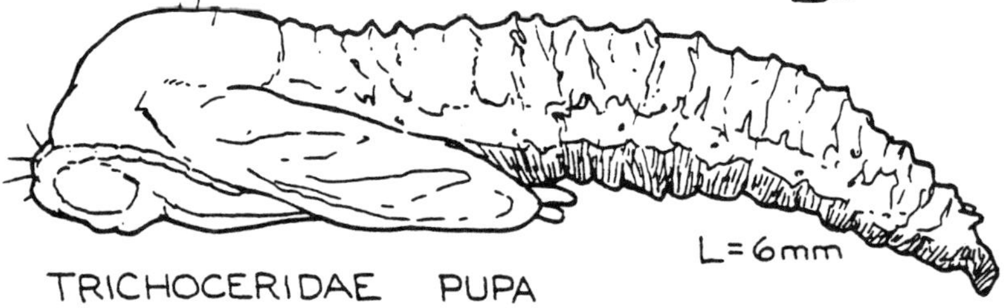

TRICHOCERIDAE PUPA L=6mm

3.33

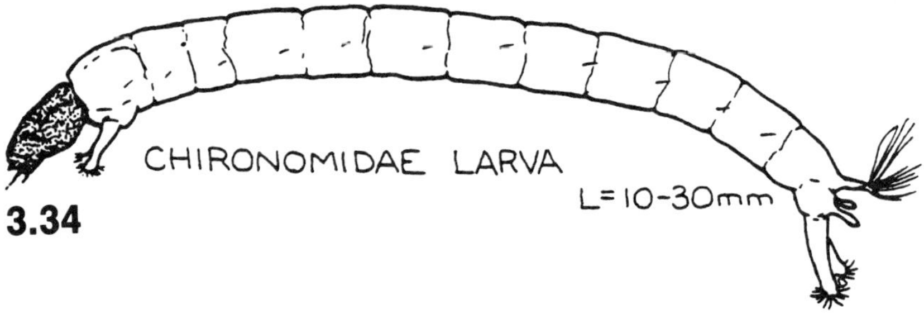

CHIRONOMIDAE LARVA L=10-30mm

3.34

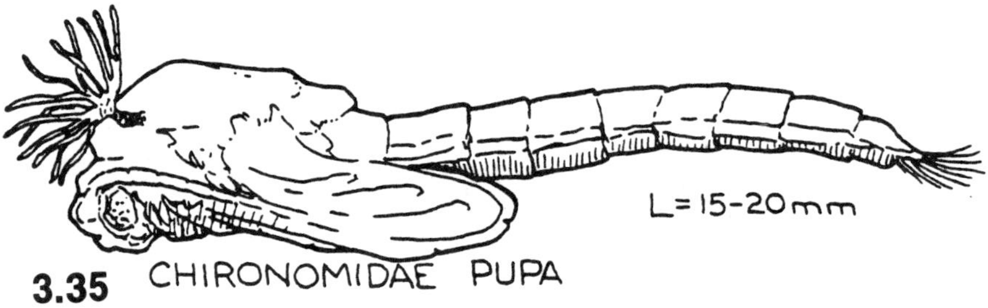

3.35 CHIRONOMIDAE PUPA L=15-20mm

IMMATURE AQUATIC FLIES ASSOCIATED WITH SUBMERGE CARRION

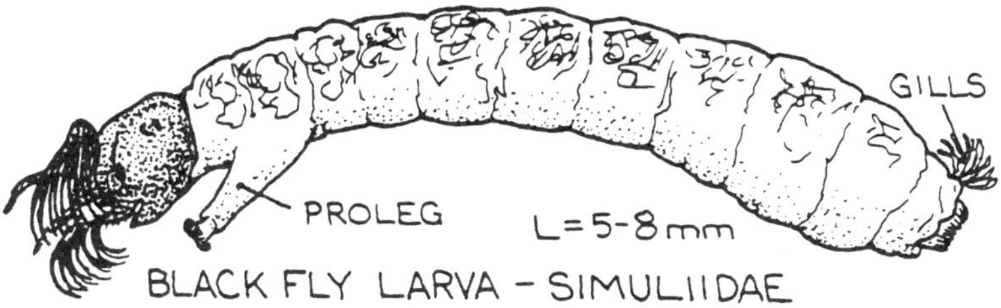

BLACK FLY LARVA – SIMULIIDAE

3.36

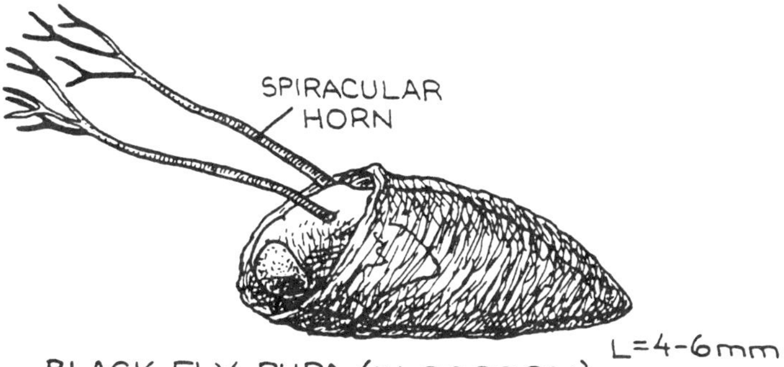

BLACK FLY PUPA (IN COCOON)
SIMULIIDAE

3.37

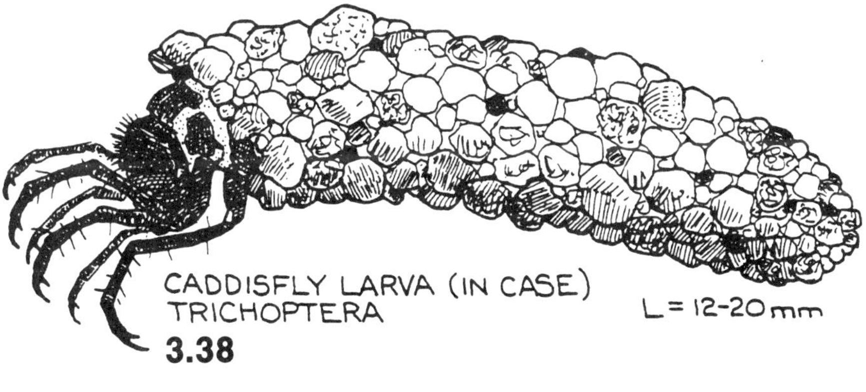

CADDISFLY LARVA (IN CASE)
TRICHOPTERA

3.38

SOME OTHER AQUATIC IMMATURE INSECTS FROM SUBMERGED CARRION

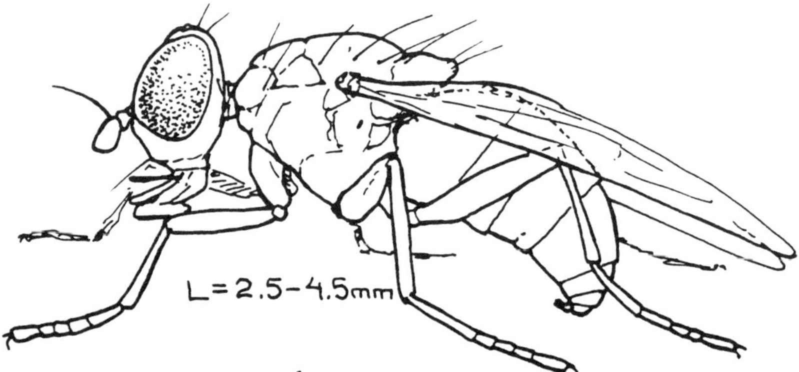

PIOPHILIDAE (Stearibia sp.)
3.39

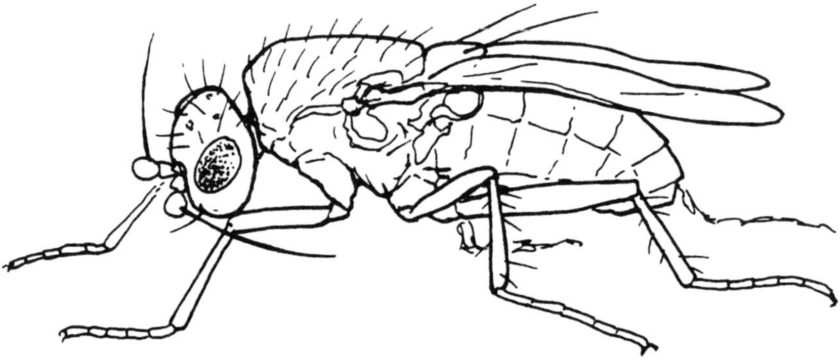

SPHAEROCERIDAE L= 3-4 mm
3.40

COMMON SMALL FLIES ASSOCIATED WITH CARRION

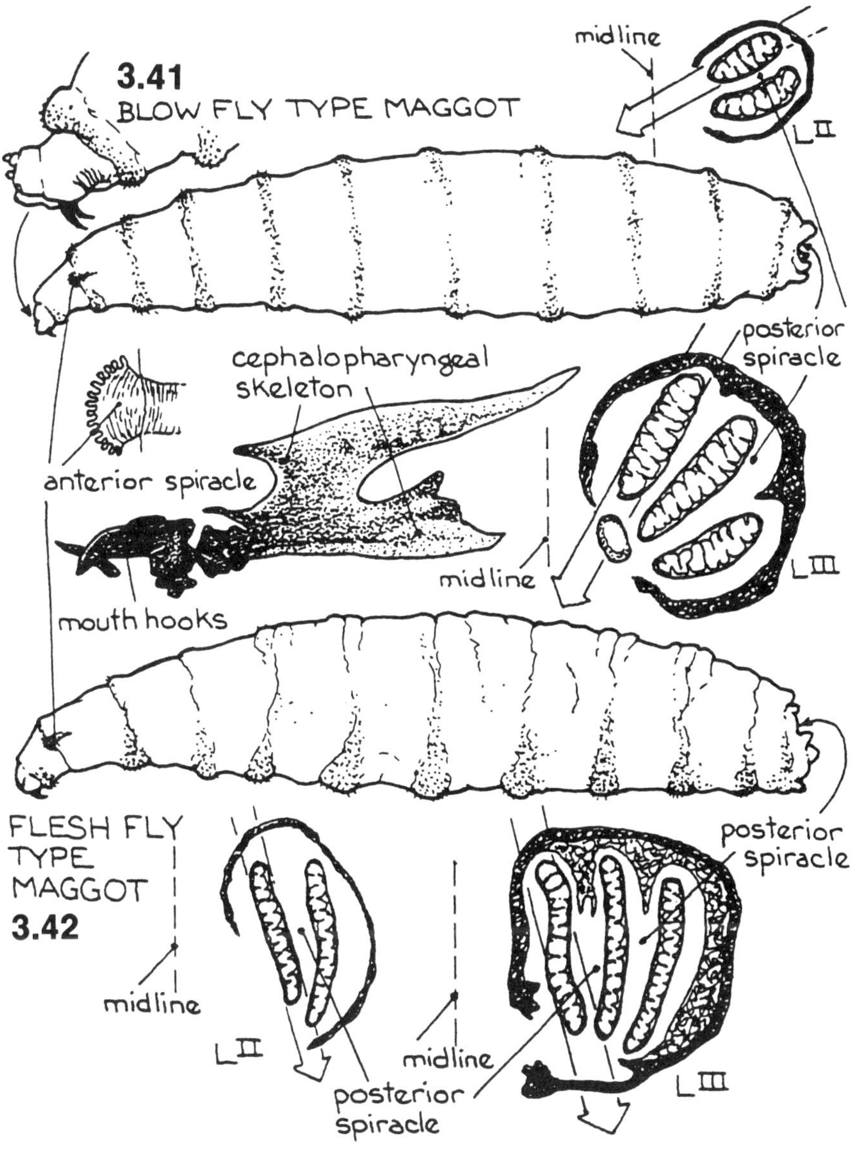

LARGER MAGGOTS ASSOCIATED WITH CARRION

3.43

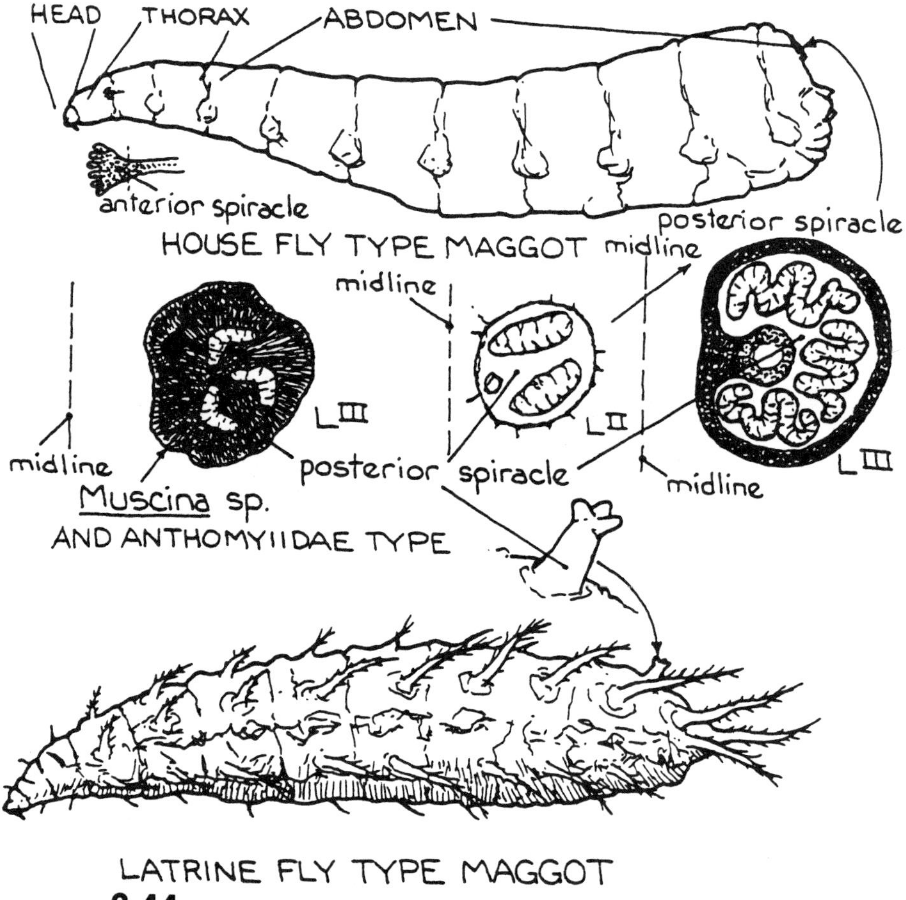

3.44

MEDIUM SIZED MAGGOTS ASSOCIATED WITH CARRION

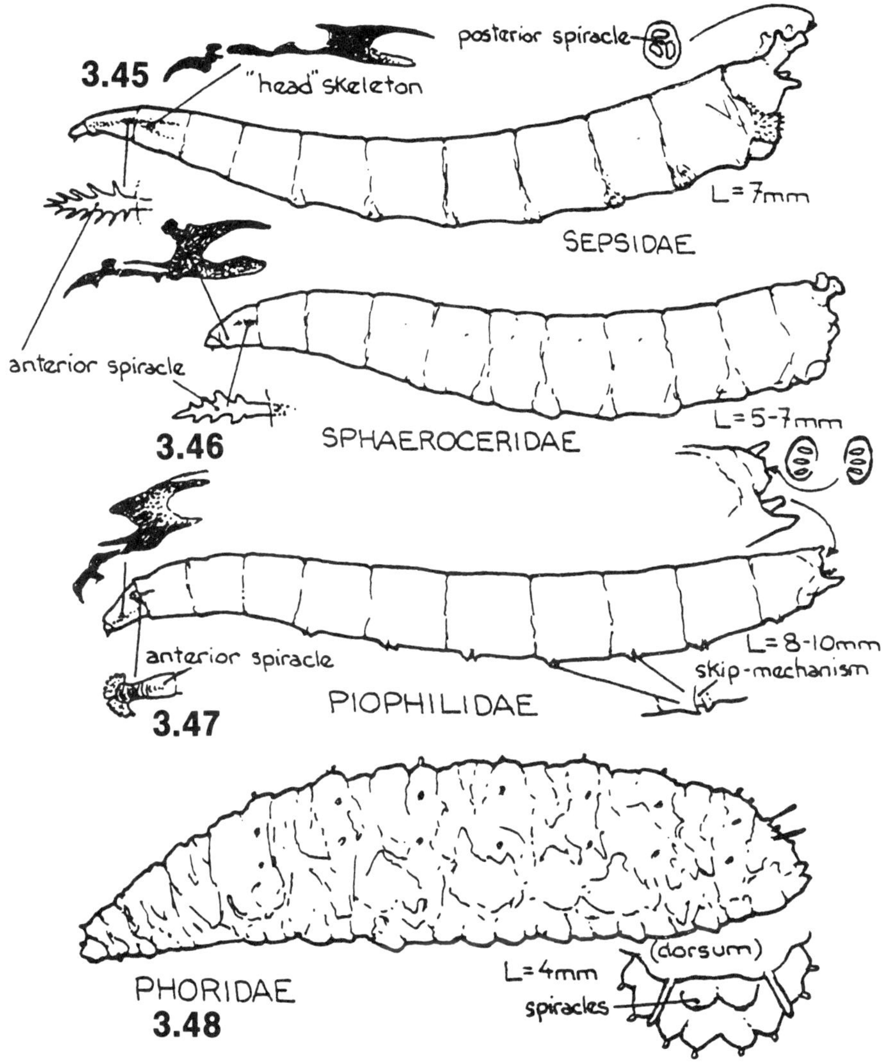

LARVAE OF SMALL FLIES ASSOCIATED WITH CARRION

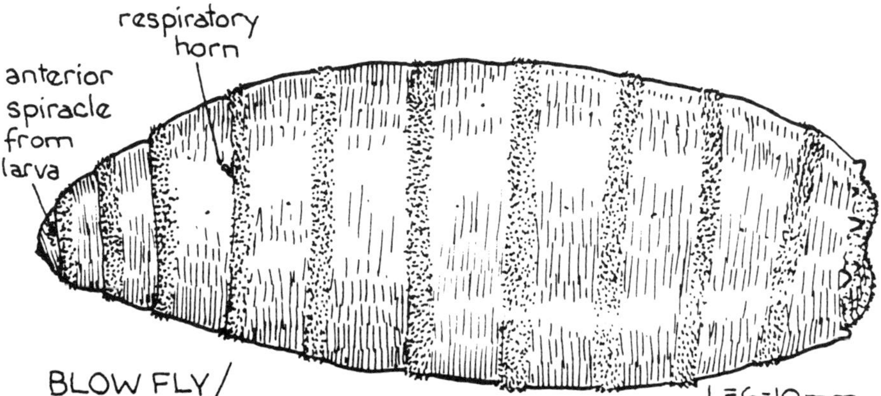

BLOW FLY/
FLESH FLY PUPARIUM
3.49

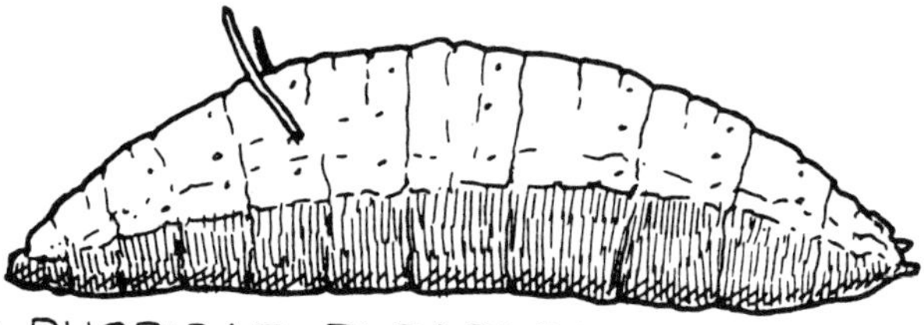

3.50 PHORIDAE PUPARIUM
Megaselia sp.
L=3mm

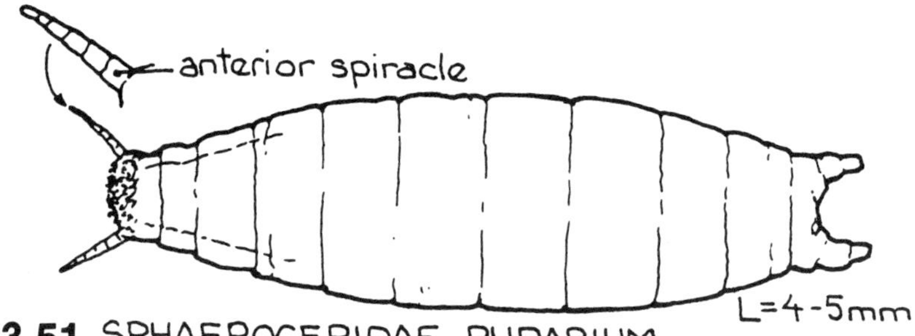

3.51 SPHAEROCERIDAE PUPARIUM
(dorsal aspect)

PUPARIA ASSOCIATED WITH CARRION

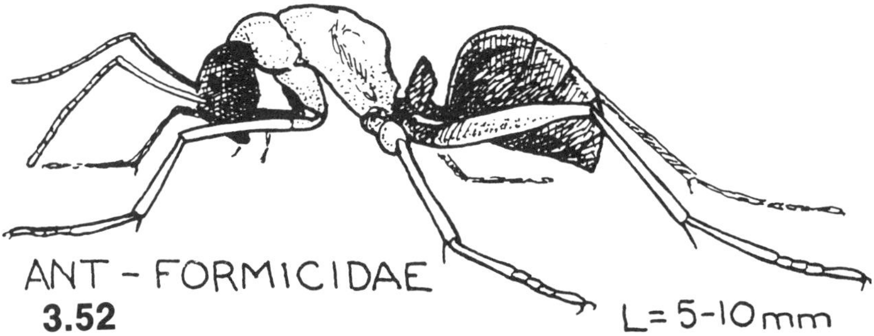

ANT - FORMICIDAE L=5-10mm

3.52

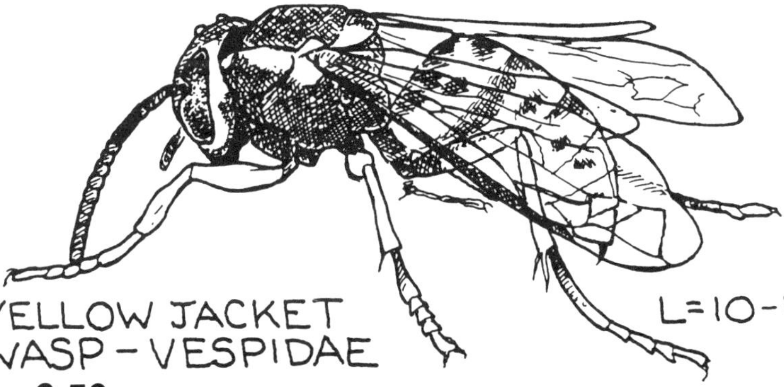

YELLOW JACKET WASP - VESPIDAE L=10-12mm

3.53

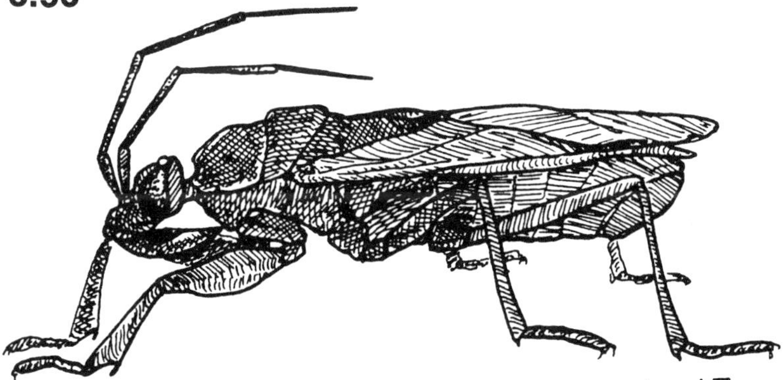

ASSASSIN BUG - REDUVIIDAE L=15mm

3.54

PREDATORY INSECTS ASSOCIATED WITH CARRION

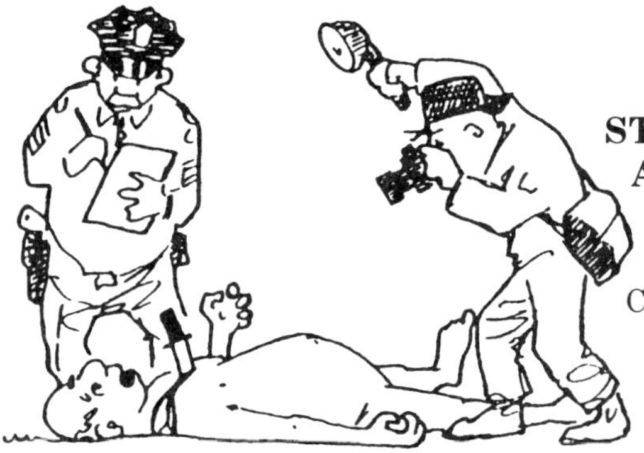

CHAPTER 4

STANDARD TECHNIQUES AND PROCEDURES AT THE DEATH SCENE[1]

C. L. Meek and C. S. Andrews

Louisiana State University and
Louisiana State Police
Baton Rouge, La

Variation in investigative terminology is often encountered when working with law enforcement agencies around the country. For purposes of this procedural guide, the term **death scene** is simply the place where the corpse is discovered. If the death appears to be by design (e.g., homicide/suicide) then it is referred to as the **crime scene**. To be more specific, it is the **primary crime scene** regardless of whether or not the death occurred at the particular site. Other sites associated with the death regardless of how distant or how many are referred to as **secondary crime scenes.**

The death scene is usually the focal point of the criminal investigation. Discoveries made at the death scene establish the direction of the subsequent investigation and will play an important role in either the indictment of or elimination of suspects. A death scene that has been processed properly can yield information that can be used to reconstruct events, link a suspect to the victim or scene, or establish the credibility of the statements made to investigators by witnesses. This information can be as basic as the identity of the victim or as subtle as the discovery of trace foreign materials that may later link the suspect to the scene.

The death scene is a place where a multiplicity of individual disciplines unite as a forensic team into a unified purpose of evidence retrieval and preservation for later interpretation and laboratory analysis. Representative members composing the forensic team may include medical examiners/coroners (hereafter referred to collectively as medical examiners), death scene investigators, meteorologists, pathologists, laboratory technicians, entomologists

---

[1] Approved for publication by the director of the Louisiana Agriculture Experiment Station as manuscript number 90-17-4395.

and anthropologists. Of course, not all of these disciplines are always represented at the death scene, nor are they involved in each case to the same extent. This is particularly true for the service disciplines (e.g., laboratory analysis technicians, anthropologists and meteorologists). However, they may become involved during the course of the investigation depending on the nature of the case and the need for verification of essential evidence.

This procedural guide addresses a wide range of subjects any of which may or may not be of general knowledge to death scene investigators, law enforcement officers and other members of the forensic team. This chapter contains information in which the typical death scene investigator and law enforcement officer have been trained fully. This information will be helpful for the forensic entomologist in understanding the role each forensic team member plays during and after the death scene investigation. This chapter will introduce the entomologist to those procedures and techniques typically used by investigation officials at the death scene. By understanding the cooperative concept of death scene investigation, the entomologist will gain an appreciation for the need to function as a member of the forensic team. By the same token, the death scene investigator should understand that the entomologist may contribute significantly to the investigative process. For example, the investigator must realize that all flies are not the same. Flies have specific behavioral and biological characteristics that an entomologist can recognize which could provide a unique piece to the investigative puzzle. This chapter will review those activities involved in: (1) identifying the probable investigative personnel; (2) securing the death scene; (3) retaining death scene witnesses; (4) conducting non-entomological investigative procedures at the crime scene and (5) conducting an autopsy by the medical examiner.

DEATH SCENE PERSONNEL

Numerous authorized personnel responsible for certain aspects of the overall investigation may be present at the death scene. These may include, but are not limited to, the following: the medical examiner or his representative, a local detective or police officer and the death scene technician. Often the **medical examiner** is designated formally as having jurisdiction over the investigative activities at the scene. However, that responsibility may be delegated to other appropriately trained individuals. The **technician** may be either a member of the county or state crime laboratory, a designated officer specifically trained in evidence collection from the local police department or a representative from the district attorney's office. It is imperative that all personnel involved in the collection of evidence conduct their investigation in a cooperative way with assigned leadership and delegated responsibilities. This coordinated effort most often yields the greatest return on the evidence available at the death scene.

DEATH SCENE SECURITY

The first officer(s) to arrive at the scene has responsibility to prevent access to the scene by onlookers, curiosity seekers and other unauthorized personnel. Improper security at the death scene may result in damage, loss or unnecessary movement of items that could render evidence inadmissable in a court of law (Fox and Cunningham 1973). If applicable, the law enforcement officer will establish a restraining line (i.e., barricade tape) around the area considered to be the "death scene." There are no set rules or regulations that predetermine the dimensions of a death scene because every situation is unique. However the critical area that composes the death scene is generally determined following the arrival of the first death investigation official(s).

The maintenance of security throughout the entire evidence gathering process at the death scene is very important. Security is lifted only after the body has been removed and the site investigation declared complete by the investigators. This period may vary from a few hours to several days. In addition, an officer will be assigned the responsibility of maintaining a list of names of individuals entering and exiting the death scene, including their times of arrival and departure.

INVESTIGATIVE PROCEDURES AT THE DEATH SCENE

Death Scene Witnesses — Often, potential witnesses are removed from the immediate area of the death scene by law enforcement officials. They also are separated from each other until they can be questioned individually in more detail by investigators who are trained in the proper techniques of interviewing death scene witnesses. Witnesses being isolated from the death scene and from each other tend to reduce collaboration and distortion of eye witness accounts. The first officer at the scene will obtain general facts from the witnesses in order to provide a brief overview for the other investigators arriving later. He does not conduct an immediate in-depth interrogation of the witnesses at that time.

Evidence Retrieval — The first officer on the scene has the responsibility to stabilize any potential "disappearing evidence," that is in imminent danger of being lost or destroyed (e.g., covering footprints and/or tire tracks with plastic sheeting to prevent damage from rainfall or extinguishing any fires that may alter or destroy evidence). Any samples of endangered evidence collected by the first officer, prior to the arrival of the investigators and death scene technicians, will be secured and labeled relative to their exact location within the death scene, time, date, case number and collector's initials. If a search warrant is required before continuing with additional processing of the death scene, then steps should be initiated to secure that document. Upon arrival of the investigators, the first officer to have arrived will provide them with a general briefing of the known facts. The investigators may then conduct a quick, post-briefing reconnaissance of the area to verify the briefing statements.

The type of scene, its location and the weather conditions determine how the scene will be processed for various kinds of evidence. In an indoor location, the corpse and other potential evidence are relatively protected from accelerated destruction due to unfavorable weather conditions, such as rainfall, direct sunlight, and extreme fluctuations in ambient temperatures. Outdoor scenes may be altered by weather conditions, catastrophic events (landslides, fire, etc.) and vertebrate scavengers. Sometimes the body may have been partially mutilated by wheeled vehicles or farm equipment prior to being discovered. Evidence inside of vehicles is, for the most part, protected, however, it may be exposed to a wider range of temperatures than other indoor or outdoor habitats. Whenever possible, outdoor death scenes should not be processed at night. Artificial lights cause glare, multiple shadows and color shift problems that obscure certain types of evidence, particularly photographs. It is best to wait for daylight in order that shadows caused by floodlights do not hide vital evidence that would be more visible in natural lighting. Such a death scene, of course, requires overnight police security.

The death scene technician's duties are to reconstruct events as they most likely happened at the scene and to identify and properly collect the different forms of evidence. The evidence items collected should corroborate the reconstruction of events, establish the credibility of statements made by witnesses, or be used to link the suspect to the scene. Sometimes it is necessary to consult with specialists from a crime laboratory (e.g., toxicologist) or a university (e.g., forensic anthropologist or entomologist) when the technician is uncertain about how useful or how much information may be gained from a certain piece of evidence. The general rule is that the technician collects anything perceived to be useful in reconstructing events or linking a suspect to a scene. Small evidence items, such as insect fauna, that might be lost or destroyed during the normal processing of the remainder of the scene can be classified as disappearing evidence and should be collected, packaged and labeled early in the investigation.

The next task in the process of collecting evidence is the dusting of selected objects for fingerprints. Better results are obtained when evidence is dusted at the scene than in the laboratory. Latent prints can be smeared or rubbed off while an object is being transported to the laboratory. The dust is applied to hard, smooth surfaced objects with a brush to make latent prints visible. The developed print is then "lifted off" of the object usually with clear plastic tape and applied to a backing material such as clear vinyl acetate squares or strips. Paper and other porous items requiring print processing are packaged and brought to the crime laboratory where they are treated with chemicals to develop the prints. Deep tire or shoe prints are preserved by making a casting of plaster or liquid epoxy.

Wet and dry blood specimens are collected on small strips of clean linen. Dried blood is collected by wetting the linen strip with a few drops of distilled water, then placing the wet strip on the blood stain. These strips are then air dried before packaging and labeling. Other body fluids, such as semen, saliva and sweat, also can be collected in this manner. Eight percent of the people in the general population are secretors. This genetic trait means that their blood type also can be detected in body fluids other than blood. For example, their blood type can be obtained from saliva left on a cigarette butt.

In the cases of homicide, the manner in which the victim was killed is important in determining the forms of evidence sought at the scene. If a firearm was used, a search is made for gunshot residue, projectiles, cartridges, cartridge casings, shotgun shell wadding and bullet holes at the scene. Sometimes it has been the forensic entomologist who has found such evidence while searching for insect pupae along the interfaces of walls and carpets.

When the murder weapon is a knife, the perpetrator sometimes is cut while struggling with the victim. Special attention should be given to the pattern of blood spatters in such cases. Pieces of the murder weapon (e.g., knife blades) may remain at the scene. For example, the blade may break off in the victim or be dropped inadvertently at the scene. Such pieces can be matched physically to a suspect's weapon. Sometimes pieces of broken fingernails or skin fragments also can be matched to that of a suspect.

The death scene will be photographed thoroughly to demonstrate the spatial relationships between the victim, items of evidence and various landmarks. The use of markers may be needed to demonstrate spatial relationship of small obscured items, especially in outdoor scenes. Wide area photographs are taken to establish the geographical relationships of the landmarks. Aerial photography may be required to demonstrate widely spaced landmarks or geographical relationships of multiple death scenes. The use of ladders and power company aerial equipment, for example, may be readily available in certain instances to assist in photographing wide areas. Sequential photographs are taken of the possible and probable entrance and exit routes to and from the death scene. When photographing certain types of evidence (e.g., shoe prints) a reference of measurement should be included in one of the close-up photographs. Several exposures of each close-up should be taken using varying lighting techniques. Sometimes small pieces of evidence items are hidden or partially hidden from view. In those situations, photographs are taken before and after moving the obstruction.

The most important object to be photographed in detail at the scene is the corpse. It is photographed from a sufficient number of angles to demonstrate the gross features, such as the position of clothing, the corpse's appendages and proximity to other objects or landmarks. The allied members of the investigating team (e.g., forensic anthropologist and entomologist) should have access to these and other photographs taken at the death scene to assist in their evaluations and determinations of the case. Other evidence items that are documented on film include the direction of dried blood flow on and around the corpse and the degree of gravitational blood pooling (lividity or livor mortis) seen under the skin surface in unclothed body areas. Any other significant trace evidence (e.g., hair and fibers) will be photographed in place and either collected prior to moving the body or protected from loss in transit (e.g., bagging of victim's hands to preserve finger prints and trace evidence). Once the body has been turned over, the underside of the corpse and the surface where the body has been laying is photographed. During the subsequent autopsy, all wounds and other abnormalities concealed by clothing will be photographed.

In recent years the use of video tape recorders has proven to be a valuable asset in recording the death scene. Even though the cost of video tape equipment is becoming more reasonable, the purchase price can be somewhat prohibitive, particularly to strained operating budgets of small police departments. In those cases, the video tape equipment may be rented from a local business or perhaps borrowed. The equipment allows the investigators to repeatedly review the death scene activities as they were conducted for disclosure of additional information not observed initially. However, the 35 mm camera continues to be the primary workhorse in documenting the death scene. Court admissable photographs must be undistorted, true and faithful reproductions of the events and/or evidence (Fox and Cunningham 1973).

A rough sketch of the death scene should be drawn in addition to the photographs. Distances and compass directions between objects at the scene must be measured accurately and noted on the sketch so a formal drawing to scale can be made later. The simplest sketch format depicts how objects at the scene would appear if viewed from a point directly above the scene. In an indoor situation, this can be expanded to include the walls by depicting them as if they had collapsed outward to lay on the same plane as the floor and with the floor still in contact with the bottom of the wall. When there are numerous items that need to be represented in a small area of the sketch, a separate sketch with expanded scale should be drawn. As the death scene is photographed, sketched and processed for evidence by the technicians, notes should be made of their observations. Some investigators record these observations on the audio portion of the video tape recordings. These written or oral notes should contain information, whenever applicable, on weather conditions, indoor temperature, light switch positions, thermostat settings and other more unusual influences that may have impacted the death scene. This may include insect activity in or around the corpse. Occasionally, there are observations that are difficult or impossible to represent in a photograph. In such cases, the death scene investigator should make a note of the information or questions for later study.

In most instances the bulk of the evidence available at the death scene will be non-entomological and will be collected by technicians trained in the standard operating procedures for securing such evidence. That is not to say that the available entomological samples taken at the death scene are not important. In fact, there are many instances in which entomological data have proven to be pivotal in the outcome of the case.

Forensic entomologists and specialist members of the investigative team must exercise caution in order not to intrude prematurely into evidence gathering by the death scene technicians. Such intrusion includes (1) casually walking about the death scene, thus inadvertently destroying footprints and/or tire tracks of the perpetrator of the crime; (2) moving items or disturbing the corpse; (3) handling items or touching surfaces that may yield latent fingerprints or (4) removing an item from the scene without specific permission of the person in charge of the investigation (Fox and Cunningham 1973). Only after a full briefing with the official in charge and receiving permission to begin sampling should the entomologist proceed with the collection and preservation of the entomological evidence. The briefing should include an overview by the entomologist of how the entomological evidence will be collected from the scene. At this time the entomologist should request special needs and requirements (if any) to gain access to the insects to insure the death scene is not inadvertently disturbed. The entomologist should inquire about the delineation of special security areas (i.e., those areas of the death scene that may not be sampled by the entomologist) if applicable. However, the entomologist must exercise caution in proceeding with the collection of faunal evidence in order not to interfere with other evidence gathering procedures that may require more time. A more detailed description of insect and arthropod sampling techniques is discussed in Chapter 5, including an extensive list of collecting equipment, supplies and preservatives.

Once the investigators release a crime scene, the judicial and analytical value of any evidence subsequently obtained from the area is lessened. In some instances, the court will not allow evidence collected under those circumstances to be introduced at the trial. The amount of information obtainable from certain types of evidence decreases rapidly as time elapses between collection at the scene and analysis in the laboratory.

PROCEDURES AT THE AUTOPSY

When the body is removed from the death scene, it is taken to a facility equipped to conduct the autopsy. Many hospitals lack the necessary equipment for conducting a crime case autopsy (e.g., independent air conditioning system, storage facility, etc.); therefore, often the body is taken to a local funeral home. At the autopsy, the death scene technician and the entomologist should brief the medical examiner as to what evidence was discovered at the death scene, if in fact the medical examiner was not present at the death scene. The technican relates the preliminary reconstruction of events believed to have occurred at the scene and advises the medical examiner of any unusual conditions present.

The technician should photograph the body again, request reference specimens and obtain inked prints of the fingers and palms. Wounds and other damage to the body are photographed.

The medical examiner collects additional evidence that is associated with the corpse *per se*. Absorbent swabs are used in the natural body openings for evidence of spermatozoa indicating sexual activity. These swabs will be analyzed for secretor blood types by the crime laboratory. Reference hair and blood specimens from the victim are obtained to compare with specimens gathered at the death scene. Blood, urine, stomach contents and certain organ tissues are obtained for toxicological examinations in the event that drugs, alcohol or poisons are involved in the crime. Whenever firearm projectiles are associated with the victim, such as bullets, bullet jackets, shotgun pellets and shotgun wadding, they are excised from the body, rinsed and packaged for subsequent analysis. Stab wounds are examined to determine the size, blade shape, direction of entry and type of instrument that may have been used. Clothing and any artifacts obtained from the body are packaged for processing later at the crime laboratory.

The entomologist should work with the medical examiner during the autopsy. The body bag used to transport the corpse from the death scene to the autopsy facility should be examined thoroughly to ascertain that no insects or insect remains are in the bag. This is particularly important if the sampling of insects is delayed until the time of autopsy. Entomologists should not hesitate to ask the medical examiner for help in searching the deeper cavities of the head and thorax for additional specimens. Likewise they should be available to assist the medical examiner or technician in conducting their respective duties. It is important for the entomologist to work quickly, yet efficiently during the autopsy. It is equally important for the entomologist to expedite an official report to the investigators following the autopsy. The report should include the identity of the insect specimens collected at the death scene and/or the autopsy. Most importantly, however, the report may provide an estimated postmortem interval regarding the victim, based on the collected species, age of the insects and other data obtained by entomologists and death scene investigators. The total responsibility of the entomologist in providing appropriate written documentation is discussed in greater detail in Chapter 9.

Occasionally, unidentified bodies are so badly damaged, mutilated or in such an advanced stage of decomposition that the medical examiner will transfer parts of the body to a forensic anthropologist. This specialist often can provide the investigator with information relating to the victim's race, sex, stature and approximate age. The anthropologist compares medical and dental records and x-rays of missing persons provided by the investigators to the bones of the victim supplied by the medical examiner.

SUMMARY

All of the aforementioned specialists composing the forensic team play important roles in solving the mysteries surrounding an unusual death or other forensic investigations. Death scene technicians, crime laboratory analysts, entomologists, botanists, and anthropologists should work as a team to glean the most information possible from the available evidence. Open communication of case facts among these individuals is often the key factor in solving a complex homicide mystery. Suspects confronted with an array of evidence that clearly describes the death scene scenario often confess, thereby saving the court systems valuable time and decreasing expense to the taxpayer.

CHAPTER 5

COLLECTION OF ENTOMOLOGICAL EVIDENCE AT THE DEATH SCENE

Neal H. Haskell and Ralph E. Williams

Purdue University, West Lafayette, IN

The proper collection and preservation of insect specimens at a death scene is imperative. Accurate determination of specimens to species level can be accomplished only if morphological characteristics are intact and preserved for study. The following list of equipment and supplies will facilitate proper collection and preservation of specimens from the death scene and during autopsy for delivery to forensic entomological laboratories in a physical condition necessary for evaluation by the entomologist. These items are adequate for collecting the insect specimens and other associated evidence, and for maintaining and preserving this evidence in good physical condition for subsequent study.

COLLECTING MATERIALS AND EQUIPMENT

1. Aerial and/or sweep insect nets (e.g., 15″ or 18″ diameter bags with 3′ handles; 15″ or 18″ collapsible nets with variable length handles).
2. Collecting vials (1 - 2 dram size) with neoprene stoppers; screw cap collecting vials (4 dram size).
3. Wide-mouth (8 oz) pharmacy bottles with screw tops.
4. Light tension larval forceps; needle point watchmakers forceps; med-fine point dissecting curved forceps; dental picks.
5. Camel hair brush (No. 2).
6. Plastic or heavy cardboard containers, 32 - 64 oz size.
7. Plastic specimen cups, screw cap lid type, 4 oz size.
8. Paper labels, heavy quality paper — for placement inside of collection container.
9. Paper labels, adhesive backed — for placement on the exterior of collection containers.

10. Dark graphite pencil (#2) — for marking paper labels. Important: Use only pencil for labeling — ink will run in fluids.
11. Hand trowel or 4-6″ core sampling tool — for soil sampling.
12. Thermometers, electronic and/or mercury.
13. Psychrometers, electronic and/or sling type — for measuring relative humidity.
14. Camera, 35 mm with macro lens and flash; dual slide/print film; video camera/recorder.
15. Paper towels and tissue paper — used in rearing of live specimens (growing a particular insect stage to adult) or cleaning thermometer probes, forceps, or other equipment after collecting.
16. Solutions for preserving specimens (uses and formulations described in Chapter 7).
17. Disposable surgical or polyethylene gloves.
18. Household deodorizer — for masking offensive odors.
19. Eye dropper pipettes — for collecting very small insects in fluids.
20. Insect aspirator, battery powered or blow type — for collecting small crawling and flying insects (Fig. 7.5).
21. Flashlight or other portable light source.
22. Measuring devices, rulers, grids, tapes, etc. — for photographic purposes and/or for obtaining measurements and distances of other evidence.
23. Berlese funnel — for extracting fauna from leaf litter, soil samples, other material. (Fig. 7.3)
24. Shipping containers — styrofoam boxes with lids; small rectangular cardboard boxes [5 × 10 × 12 cm (2 × 4 × 5 in)] or cardboard and metal screw cap cylinders [7 × 15 cm (3 × 6 in)]; shock absorbing type packing.
25. Chemical ice (not dry ice) — for cooling and maintaining specimens collected alive.
26. Death Scene Procedures Checklist (see inside back cover).
27. Log Book — for recording location, scene data, date, etc.
28. Data sheet — Death Scene Case Study Form (see Chapter 9 and Appendix A.2).
29. Data sheet — Specimen Deposition and Identification Record (see Chapter 9 and Appendix A.4).

Much of the listed equipment may be purchased from biological supply and scientific equipment companies.

PROCEDURES

The forensic entomologist (or crime scene technician) will encounter a wide variety of habitats when investigating various death scenes. Examples of habitats can include open, grassy areas; deciduous forests; mountainous tundra; alpine slopes and meadows; high plains deserts; swamps; drainage and irrigation ditches; river banks; lakes and ponds; darkened closed rooms in abandoned buildings; shallow burials; or closed trunks of automobiles. The following procedures are generally applicable to any geographic area or habitat. Similar, less detailed treatments of this subject have been done by Lord and Burger (1983) and by Smith (1986). Specific techniques described for habitats requiring additional or special collecting procedures are discussed in Chapter 6.

COLLECTION OF DATA AT THE DEATH SCENE

If the forensic entomologist is called upon to visit the death scene, there are restrictions that may be imposed on the immediate sampling of insects. In any investigation where deliberate death is suspected, protection of the scene to preserve all possible evidence is paramount. The sampling of insects, mites and worms can result in some unavoidable disturbance to the remains, so utmost caution and care is required.

The entomologist will work with either the primary investigator or one of the subordinates. This individual should be contacted by the entomologist immediately upon arriving at the scene. Also, it would be highly beneficial for the entomologist to advise the primary investigator of a plan for collecting entomological evidence. As noted in Chapter 4, security for the area is critical and is the responsibilty of the primary investigator. Therefore, the entomologist must be patient while other portions of the scene processing are underway, waiting until approval is given before entering the restricted area. The investigating entomologist can use the time spent waiting to note the general habitat, weather conditions, and location of the body (if visible), to make a mental note of the kinds of insects anticipated to be found in the scene habitat, and to observe the overall aspects of the surrounding area. These impressions should be recorded in writing (Death Scene Case Study Form) or dictated on cassette tape for later study.

When authorization has been given to enter the secured death scene area, direct communication with the crime scene technician is essential. The crime scene technician will brief the forensic entomologist as to the circumstances surrounding the death scene and may provide valuable suggestions regarding past events at the scene. Technicians are usually very interested in the entomological aspects of the investigation and may have suggestions as to additional collecting techniques, or other locations within the secured area for collecting insect evidence, or they may even help with the entomological collection itself. Once this information is obtained from the technician, the forensic entomologist can begin the entomological portion of the investigation.

The sequence of the entomological investigation of the case at the death scene can be broken down into five major stages:

1. Visual observations and notations of the scene.
2. Initiation of climatological data collection at the scene.
3. Collection of specimens from the body before body removal.
4. Collection of specimens from the surrounding area [up to 20 ft. (6 m) from the body] before removal of the remains.
5. Collection of specimens from directly under and in close proximity to the remains [3 ft (1 m) or less] after the body has been removed.

Observations and notations of insect activities at the death scene by the entomologist may prove useful to the investigation due to the fact that the forensic entomologist is trained in a different science than the crime scene investigators. The entomologist may observe elements of the scene which the crime scene investigators may ignore. The crime scene investigator has a tremendous task and responsibility at hand when attempting to clarify the circumstances and to collect evidence related to a death scene. A wide diversity of evidence may be present and many different procedures may be required to recover this evidence. Often, only one person is available to work the scene, which means there could be more work than one person can physically handle within the time constraints allowed the investigator. Given the pressures of the public press, police administrators, or public officials for immediate answers from the death scene, important but obscure evidence may be passed over and not collected or noted. Therefore, the personal observations and notes of the forensic entomologist may provide valuable information for the overall death scene investigation and important substantiating data for the entomological evidence being evaluated from the scene.

DISAPPEARING EVIDENCE

Some insects can be classified as "disappearing evidence." This kind of evidence is that which is temporary, transient, or unstable. Examples include frozen or fluid substances that can evaporate, items which are burning, and fingerprints, footprints and tireprints in snow or mud. Flies, maggots and beetles also can be disappearing evidence. Adult flies on the corpse will fly off when the body is approached. Beetles and maggots beneath the corpse will bury themselves quickly when the body is removed. If these insects are not collected at the time they are first observed, then a record of the kinds of insects (if known) or, at the least, a description of the insects and their approximate numbers should be made.

VISUAL OBSERVATIONS AND NOTATIONS OF THE SCENE

Visual observation and notation of the insects active in and around the remains can be of value even before the remains are examined. The forensic entomologist may accomplish this by:

1. Entering records on the "Death Scene Case Study Form," briefly discussed in Chapter 9 and formatted in Appendix A.
2. Approximating the number and kinds of flying and crawling insects.
3. Noting locations of major insect infestations associated with the body or surrounding areas. These infestations may be eggs, larval, pupal, or adult stages in any combination or alone.
4. Noting immature stages of particular adult insects observed. These stages can include eggs, larvae, pupae, empty (eclosed) pupal cases, cast larval skins, fecal material, and exit holes or feeding marks on the remains.
5. Noting any insect predation such as beetles (e.g. silphids, staphylinids), ants and wasps (e.g. formicids, vespids) or insect parasites (e.g. ichneumonid and chalcid wasps).
6. Noting the exact position of the body: including the compass direction of the main axis, position of the body extremities, and position of the head and face; noting of which body parts are in

contact with substrate; noting where there would be sunlight and shade in the immediate area during a normal daylight cycle.

7. Noting insect activity within 10 - 20 feet (3 - 6 m) of the body. Observing flying, resting, or crawling insect adults and larvae or pupae within this proximity to the body.

8. Noting any unusual naturally occurring, manmade, or scavenger-caused phenomenon which could alter the environmental effects on the body (e.g., trauma or multilation of the body, burning, covering or enclosing of the body, burial, movement, or dismemberment).

Written notations of the scene can be augmented with photographs and drawn sketches. Closeup detailed photos of different stages of insects found may provide important clues in analyzing the entomological information at the scene. Some closeup photos should be taken prior to collecting the insects. It may be possible for the forensic entomologist to have the crime scene technician take the needed photos. Dual slide/print film can be used in recording entomological evidence. This film provides an uncut negative strip which demonstrates the filming sequence and facilitates duplication of photos. However, this film may not have as high a resolution as slide or print film. A record of the exact time when each photograph was shot also should be made.

The scene also may be recorded with a video camera. While this is an excellent procedure for recording overall elements of the scene and observing entomological information from a gross perspective, the fine details and closeups required in entomological investigations usually cannot be viewed on a video tape. The video camera should not be used as the only source for recording the details of the death scene. The 35 mm camera, with a macro lens, is the best tool to record details of entomological evidence. If the scene processing team makes a video record, the entomologist should request a copy.

COLLECTING CLIMATOLOGICAL AND TEMPERATURE DATA

Climatological or weather data are critical when estimating the PMI by using entomological evidence. The time required for insects to undergo their life cycle development is determined largely by the temperature and relative humidity associated with those species present in a particular environment. Other climatological conditions (e.g., rainfall, full sun, snowcover, fog) also may influence the developmental rates or carrion feeding habits of the insects found at death scenes. Therefore, the forensic entomologist should develop a working knowledge of climatology. The proper collection and interpretation of these climatological data by the investigating forensic entomologist are essential to the investigation when attempting to determine the PMI.

Once notations of the scene have been recorded, the collection of the temperatures and relative humidity can begin. There are several temperature readings which should be taken in association with the death scene:

1. Ambient air temperature can be evaluated by taking readings at 1 ft and 4 ft (0.3 to 1.3 m) heights in close proximity to the body.
2. Ground surface temperatures can be obtained by placing the thermometer on the ground immediately above any surface ground cover.
3. Body surface temperatures should be obtained by placing the thermometer on the skin surface.
4. Under-body interface temperatures can be obtained by sliding the thermometer between the body and ground surface.
5. Maggot mass temperatures can be obtained by inserting the thermometer into the center of the maggot mass.
6. Soil temperatures should be taken immediately following body removal at a ground point which was under the corpse prior to removal. Also, take soil temperatures at a second point 3 ft. - 6 ft. (1 - 2m) from where the body lay. These should be recorded from 3 levels: directly under any ground cover (grass, leaves, etc.); at a soil depth of 4 in (10 cm); and at a soil depth of 8 in (20 cm).

NOTE: The direct rays of the sun should not be allowed to shine on the thermometer sensing element. Radiant heat from the sun will cause readings far in excess of the true environmental temperatures. SHADE THE THERMOMETER FROM DIRECT SUNLIGHT WHEN TAKING ALL TEMPERATURES.

Estimate the duration of exposure of the corpse to direct sunlight, broken sunlight and shade for the total daylight hours. This can be accomplished by observing overhead and surrounding vegetation (shrubs, trees, buildings, etc.) or the location of windows and noting their compass direction relative to the position of the remains. If there is any question as to this relationship, observation of the site periodically throughout a sunny day will provide additional information. When direct sunlight is shining on the body, the external temperature and some internal temperatures of the body, close to the surface, will be higher than when the corpse is shaded. These higher temperatures will be recorded even if the thermometer bulb is shaded and will reflect accurate temperature readings for the body.

Relative humidity can be obtained at the scene through the use of a sling psychrometer (wet and dry bulb) or a battery powered psychrometer. From these data, dew points and dew wettings of the remains can be estimated. Under certain conditions, the wetting attributed to dewfall can be of greater cooling and wetting influence than those resulting from rain or snowfall. Even though there has been no precipitation recorded, a body may have been coated with surface moisture daily. Air movement and cooling from dew evaporation (evaporative cooling) can reduce corpse temperatures below ambient levels.

When insect collecting is completed at the scene, weather data for the time period in question should be obtained. This time period should extend from 1 - 2 weeks prior to a rough estimate of when death occurred to the date when the deceased was last seen alive to 3 - 5 days past the time the body was discovered. Weather data retrieval can be accomplished by contacting the nearest National Weather Service (NWS) Station or other climatological data gathering agency for climatological data retrieval. Locating the closest National Weather Service recording station or other climatological data gathering agency to a death scene is necessary in the proper evaluation of climatological data from that death scene. These offices have records of other data collection sites in each state, and can indicate which station is in closest proximity to the death scene. Some NWS stations (including forest service firetowers and airport facilities) can give extensive weather data including hourly temperatures, humidity, extent of cloud cover, precipitation, and wind speed and direction. In addition, soil temperatures, water temperatures, river stages, tidal swings, soil moisture conditions, and evaporation rates may also be obtained from the first order NWS stations. All data is eventually sent to NOAA, National Climatic Data Center, Asheville, NC, 28801-2696 and can be obtained in the form of published recorded material for court documentation. Other stations will have only daily maximum, minimum temperatures and total precipitation, and these data may be all that are necessary for the entomologist to make the appropriate time interval evaluation.

Finally, it is important to conduct periodic temperature observations (3 to 4 readings over a 24 h period for 3 or 4 days, particularly during the times of temperature maximums and minimums) at the death scene, after the body has been removed, to assess the correlation of the site temperatures with those of the closest NWS station data. Great differences can exist even in short spatial distances, and the possibility for this type of error must be considered. In a California case where a body was found next to a river, extensive periodic temperature data recovery was conducted at the site. It was found that the site temperatures were ca. 11F (6C) lower than the same hourly temperatures recorded from the NWS station at the airport. By making periodic visits daily (including times after dark) to the site over several days, accurate interpolations can be made between NWS station and death site. If the NWS station is recording data on an hourly basis, an even greater degree of accuracy can be achieved by recording coincidental death scene readings.

COLLECTION OF SPECIMENS BEFORE BODY REMOVAL

Proper collection and preservation techniques, specimen labeling, and data recording are necessary for entomological data to be accurately evaluated. The importance of accurate entomological data is essential for entry into the court record when the case is tried.

A passive technique (not requiring much effort on the part of the investigator) for very efficient sampling of blow flies and other flying insects was used successfully in a faunistic insect study of human remains in Knoxville, Tennessee (Haskell, unpub. data). It was found that when commercially manufactured sticky traps (using a slow drying adhesive substance similar to that coating fly paper) were placed approximately 3 feet (1 m) from human remains, large numbers of flying insects were collected on the angled surfaces. The traps [approximately 9 inch (23 cm) × 10 inch (25 cm)] are made from wax coated cardboard with an inverted "V" or pup tent configuration set approximately at a 60° angle with sticky material on both exposed sides (Fig. 5.1). Four wooden, spring type clothespins are clipped to the bottom corners for ground clearance and the trap placed in an upright position on the ground surface. The pilot study, conducted in 1989, showed that blow flies numbering approximately 60 to 80 individuals were collected after several minutes, whereas

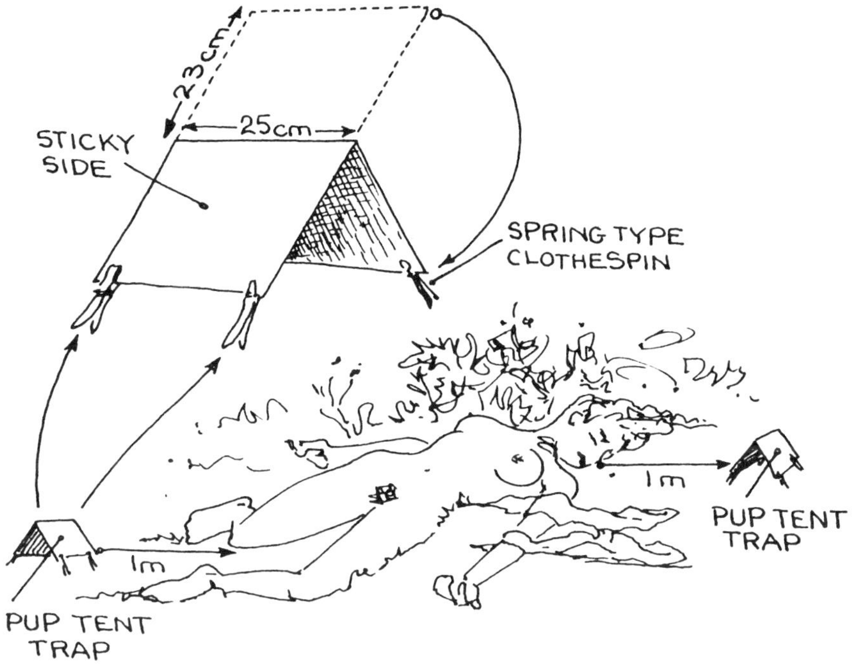

5.1 PUP TENT FLY TRAP

collecting with the aerial net yielded numbers of only 1 to 15 individuals in different samples. The traps were oriented with one surface facing the corpse and the other surface facing away. Adult blow flies and other insects can be captured when they are disturbed and fly off the carrion or when they fly off the carrion to periodically rest on nearby surfaces. The sticky traps can be labeled as to hour, date and location and gently folded back, so the sticky material is on the inside, and returned to the lab. Care must be taken not to smash the traps so that the insects will be intact for identification. Once at the lab, the specimens can be individually removed from the traps, placed into screwcap vials, washed in paint thinner or xylene, and then placed into alcohol vials and labeled for later identification.

When using sticky traps at a death scene, one or more sticky traps can be placed around the remains at varying distances [anywhere from 1 foot (0.3 m) to 12 feet (4 m)]. If only one trap is used, 2 - 3 feet (0.7 - 1.0 m) is an adequate distance. The traps can be set out before the remains are examined closely; once the remains are approached, insects which are disturbed and leave the body may fly to the surface of the trap. The traps come in several colors, but the most successful traps used in the Tennessee study were white and contained no attractant odors. Sticky trap placement at the initiation of the scene processing may lend valuable insect data concerning flying insects which may otherwise be lost once movement and recovery of the body commences.

An aerial insect net is very useful during this stage of the entomological investigation for collecting fast flying and fast crawling adult flies. If flying insects are present over the body, the aerial sweeping technique for collecting can be employed. Flying insects associated with carrion are strong, fast fliers, thus, netting of specimens requires some practice and experience. This netting technique uses several rapid, back and forth sweeping motions of the net (6 - 10 sweeps) with reversal of the opening of the net 180° on each pass (Fig. 5.2). On the last pass, the opened portion of the net is brought up to about chest level with rotation of the opening 180°. This causes the netting material to be folded over the top edge of the large net ring opening, thus trapping the insects in the net bag. The net then can be placed on the ground so that two hands can be used for trapping and confining the specimens in the tailing end of the net (Fig. 5.2, #6). The end of the net, with insects inside, can then be placed into a wide mouth killing jar, which is then capped. The killing jar should contain a few cottonballs soaked with fresh ethyl acetate. This will kill the insects after a few minutes (2 to 5 minutes is usually adequate). Following immobilization, the insects can be transferred into vials of 75% ETOH by placing a small funnel into the vial and carefully dumping the contents of the net into the funnel. Some of the specimens may be placed into a dry vial for direct pinning. If stored in this dry condition however, the insects must be processed in a few hours because excessive moisture on the insects (condensation arising from within the closed vial) can promote mold growth which will quickly damage or destroy the specimens. Care should be taken that all insects are shaken out of the net and into the vial. Many insects have spurs or claws that can catch on the netting material and hinder their falling free. Sweepnetting procedures should be repeated three to four times to ensure a representative sample of all of the flying insects present.

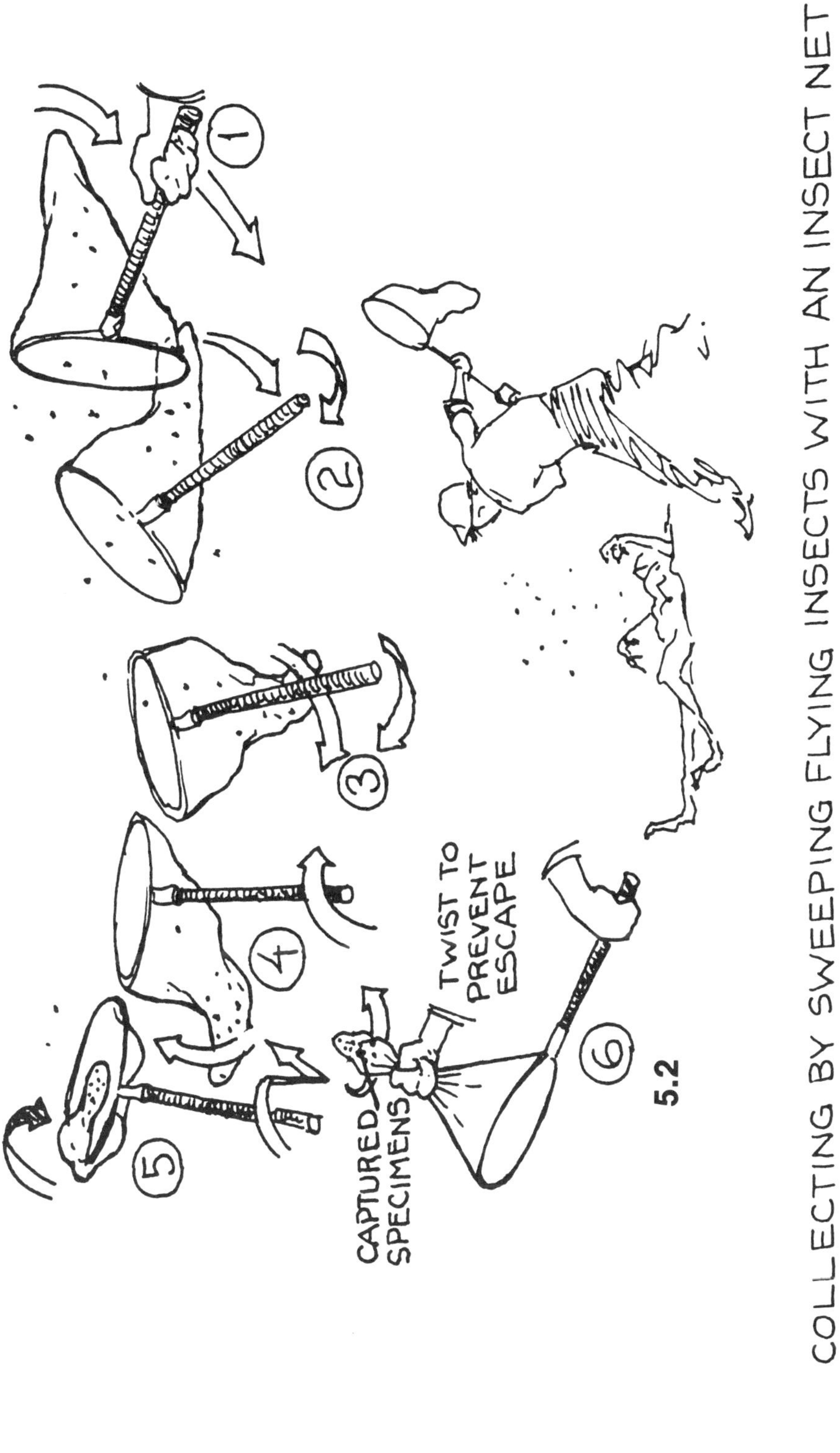
1
2
3
4
5
6
TWIST TO PREVENT ESCAPE
CAPTURED SPECIMENS
5.2
COLLECTING BY SWEEPING FLYING INSECTS WITH AN INSECT NET

Many ground crawling adults can be collected with forceps or fingers. Preservation should be conducted in the same manner as with the adult flying insects. These insects may include some of the beetles (Coleoptera), ants, bees and wasps (Hymenoptera), true bugs (Hemiptera), springtails (Collembola), and newly emerged flies (Diptera).

It is very important to put a label into each collection vial immediately. The sample number, hour, date, case number, and county/city should be included on each label. Sample labels are found in Appendix A.3. It is also important that these labels be written in pencil, NOT INK! Labels made with a graphite pencil will not be affected by solutions used to preserve specimens. Duplicate labels also need to be made and affixed to the outside of the vial. This double labeling procedure should be a standard forensic entomological practice. This is done to insure that data will not be lost due to external labels becoming loose and falling off. Also, when working with specimens, the internal label will be placed in the examination container that contains a portion of the specimens from the vial while the remaining specimens in the collection vial are still properly labeled, thus reducing the chances for mixing similar appearing samples. The vial sample number and the data pertaining to where the sample was collected should then be recorded on the "Specimen Disposition and Identification Record" and also noted on the "Death Scene Case Study Form." (See Chapter 9 and Appendices A.2 and A.4).

The sweepnetting technique described above also can be used to sample insects upon vegetation in the area surrounding the body. Many of the Diptera (flies) associated with carrion rest on nearby plants or grasses. By sweeping the vegetation within 10 - 20 feet (4 - 6 m) of the corpse, some of these adult insects may be collected. The aerial net can be used for this survey, but if the foliage is too thick or dense, or the plants are woody and stiff, the light weight material of the aerial net will snag and tear. Instead, heavy bagged sweep nets (such as those used by agricultural insect specialists for surveying insects on field crops) can be employed for collecting from such foliage.

Once the surrounding area has been processed, collection of specimens from the body can begin. The person having overall authority of the scene must give permission to approach the body although this should have been done prior to collecting around and over the corpse. The appropriate route to physically approach the corpse should be identified: many times a specific route or path is designated for movement into and out of the confines of a death scene to minimize destruction of evidence.

When the collection of insects is made before body bagging and removal, it is extremely important that nothing be moved or taken from the corpse except the insects which are on the surface and clearly visible. DO NOT DISTURB ANY PORTION OF THE CLOTHING, THE IMMEDIATE AREA AROUND THE BODY, OR THE BODY ITSELF. The entomologist should collect only those stages of insects which can be seen readily on the body, and use extreme caution when using forceps or other tools for collecting. Postmortem artifacts inflicted inadvertently while collecting specimens may be misleading and can cause needless questions to be answered. A thorough examination of the clothing and body bag (some insects may crawl off the body during transportation) can be conducted at the time of the autopsy for insects not collected at the scene. The body bag, too, must be examined for previous insect infestations before the remains are placed inside. Some agencies require new body bags for each case to eliminate the question of residual evidence from previous cases. Others may not have that requirement or the finances to permit it. In either case, verify the absence of insects in the body bag before use.

Eggs and a mixed-size sample of larvae (several hundred) should be collected and preserved in one of the specified perservatives (Kahle's is preferred) listed in Chapter 7. The largest larvae observed should be collected and, if numbers permit, several dozen of these should be placed into the solution. A second portion (2-3 dozen) of these large larvae should be placed alive into plastic specimen cups containing some damp paper toweling, or placed directly into the maggot rearing cups previously prepared (described in Chapter 7 — Procedures in the Entomology Laboratory). Representative samples of these larger larvae (several dozen individuals) are to be reared to adults. It is important that rearing of larvae to adults be undertaken to facilitate and confirm larval identifications. This procedure should be repeated for some smaller larvae or eggs, placing a portion of the specimens into preservative and also saving a portion alive for rearing.

Areas of the body where concentrated insect activity most likely will be encountered during the earlier stages of infestation are the nasal openings, ears, mouth, eyes and traumatized areas of the body (cuts, gunshot wounds, and blunt force injuries where the skin is broken). Skin creases of the neck also may contain egg masses. Egg masses also will be found along the hair line and close to natural body openings. Wounds may have both egg masses and larvae associated with them, and exposed genital and anal areas may contain egg masses and larvae, especially if these areas were traumatized prior to or after death.

As stated earlier, thorough examination of the clothing and the body for entomological evidence should be conducted during the autopsy and not at the scene so that physical evidence relevant to the medical examiner's or forensic pathologist's portion of the investigation will not be disturbed or destroyed. At the death scene, the forensic entomologist should concentrate on collecting from around and under the corpse, while collecting from the body should be superficial. The forensic entomologist should try to collect all apparent insect stages, but without causing postmortem artifacts on the body.

COLLECTION OF SPECIMENS AFTER BODY REMOVAL

In cases where bodies are outdoors and heavily infested, many insect adults, larvae and pupae as well as other arthropods will remain on the ground after the body is removed. The procedures described above should be followed for each of the different insect stages seen follwing removal of the body. A number of specimens of each immature stage should be collected and preserved, while a second sample should be collected alive for rearing.

Litter samples (e.g., leaves, grass, bark, humus) or any material on the ground suface close to or under the corpse should be collected and labeled. Many of the carrion feeding insects will shelter or hide in this material close to the body; therefore, this too can be examined for faunal evidence. Collect handfuls of the litter down to the exposed soil, particularly litter very close to the ground surface. This material can be placed into two-quart cardboard or plastic containers for subsequent examination in the laboratory.

Soil samples should be approximately a 4 inch cube or core (100 cc) of material from areas associated with different body regions (head, torso, extremities). Soil samples of this size fit well into 2 pint (1 liter) or slightly larger cardboard (cylindrical ice cream type) or plastic containers. Soil samples (about six in all) should be taken from under, adjacent to, and up to 3 ft (1 m) from the body, noting the origin of each sample in reference to the position of the body. These samples should be labelled as were the insect vials.

CHAPTER 6

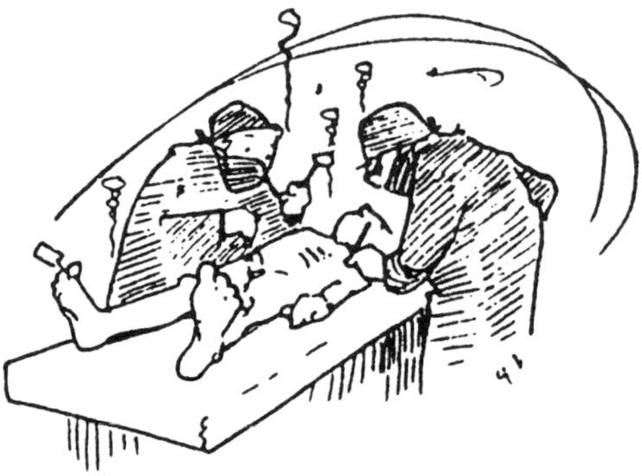

ENTOMOLOGICAL COLLECTION TECHNIQUES AT AUTOPSY AND FOR SPECIFIC ENVIRONMENTS

Neal H. Haskell

Purdue University, West Lafayette, IN

Specific collecting techniques and procedures in different environments will enhance the recovery of insect specimens. Specialized techniques have been developed for collection of insects at autopsy, insects found in burials, insects in enclosed structures, and insects found in aquatic environments. Although these are similar to general collecting procedures (Chapter 5), there are instances when an added step is necessary or an additional observation (place to look for specimens) must be made. The following suggestions for collecting at the autopsy or other types of environments will add reliability to the entomological evidence.

COLLECTION OF ENTOMOLOGICAL SPECIMENS DURING AUTOPSY

It is best for someone experienced in collecting insect evidence to be involved at each step of the investigation to make use of the full potential of entomological evidence. However, if this is not done before the remains are moved, then the best opportunity for making proper collections of insect evidence is during the autopsy. Ideally, the entomologist should be present during an autopsy involving the collection of entomological evidence. However, if circumstances prevent this, the medical examiner/coroner should be able to gather the necessary entomological evidence by following the procedures outlined here. If a certain insect life stage critical to determination is not found during the autopsy, the investigator or entomologist can return to the scene (even after several hours) where, most likely, the missing insect life stages will be found.

Collection of insect evidence at the autopsy is done in conjunction with the forensic pathologist performing the autopsy. The entomologist should tell the pathologist the proper entomological evidence collection needs, and inform the pathologist prior to taking any action. This will preclude any later problems with the validity of evidence and prevent postmortem artifacts on the body.

Most likely the corpse will be enclosed in a "body bag" (zippered plastic bag) when it arrives from the death scene, particularly if the remains are in a state of advanced decomposition. In cases where the corpse is heavily infested with insects, the outer surface of the body bag may have larvae or adults on it. This evidence should not be overlooked; these insects should be collected and labeled using procedures described previously. Once the body bag is opened, the inner surfaces of the bag should be examined for insects. The insects may have crawled from the body in transit due to changes in temperatures and now may be found on the inside surfaces of the bag. The entomologist should collect and label these insects with notations as to which area of the body (head, torso, extremities) the insects were near.

Many times remains are stored for some time in coolers or refrigeration units prior to autopsy. This period may range from several hours to several days. Therefore, notation should be made of both the total time the body was cooling and the temperature of the chamber along with any other temperature information regarding the movement of the body to the morgue (monitoring temperatures of the vehicle body compartment periodically on long transits). Also, the temperature of the maggot mass should be recorded when the body is removed from the cooler. There may be little or no effect of the lower temperatures on insect development if the maggot mass was well established prior to the body being placed in the cooler. Maggot mass temperatures may be in the 100F (27F - 37C) range even if the temperatures in the cooler are in the 30'sF (2 - 4C).

After the corpse has been taken from the body bag and placed on the autopsy table, an external examination is made. The entomologist and the pathologist can assist one another in their collection of evidence at this stage of the autopsy. If the body is clothed, a complete and detailed examination of the clothing is essential and may yield a variety of stages and kinds of insects. Folds in the clothing where eggs, larvae, pupae, or adults may be sheltering should be opened gently. Moist patches on clothing are good areas to search; fly eggs are often laid in moist areas. The collection of seeds and other plant materials should be conducted at this time for analysis by a forensic botanist. In addition to the plant parts as evidence, certain live stages of insects may be found in stems or seeds of plants which may give clues as to habitat, time, or geographic aspects in the case.

The bagging of the hands of the corpse in paper or plastic sacks taped to the wrists should be standard practice. This is done to preserve foreign trace evidence such as skin fragments or hair adhering to the hands. Inspection of these hand bags after their removal at autopsy is necessary as insects infesting hand wounds may crawl off the remains during transit.

After the clothing has been examined and removed, the areas of the body where concentrations of insect activity are found should be noted and photographed with a macro lens to show the extent of the infested area and insect composition. Collection of representative samples from each major site where insects are found should be performed using the collecting techniques described earlier. If fly larvae are present, some should be kept alive and some preserved. On fresh bodies, the face is the most likely area to have insect activity. Flies in particular will seek the moisture and shelter of external openings (e.g., nostrils, mouth, eyes) for deposition of their eggs. The genital or rectal areas, if exposed, will sometimes provide shelter and moisture for attracting egg laying (ovipositing) female flies, especially if those areas have been traumatized or soiled with excretions. However, the urogenital-anal area is generally not a primary site for oviposition. Traumatized areas of the limbs and torso, where breaks in the skin occur, may also contain insects. For example, insect infestations on the hands or forearms may suggest that the victim had sustained defensive wounds. This may not always be the case, however, and caution should be exercised when drawing this conclusion.

Very small arthropods such as fleas, ticks, mites, lice, or nits (Figs. 6.1, 6.3) may be present on fresh corpses and clothing from fresh corpses. Many may be attempting to leave the cooling body and therefore can be found in the clothing or may even hitch a ride on the examiner. Also, other small ectoparasites may be present on the corpse itself; thus, it is important to examine the hair close to the scalp for the presence of nits (lice eggs) (Fig. 6.2). The eye lashes or sebaceous gland areas of the face may harbor follicle mites, *Demodex* sp. (Acari: Demodicidae) (Fig. 6.4) A few lashes are typically plucked and examined microscopically if the presence of this mite is suspected. Estimation of time intervals may be determined based on whether or not these arthropods are still alive.

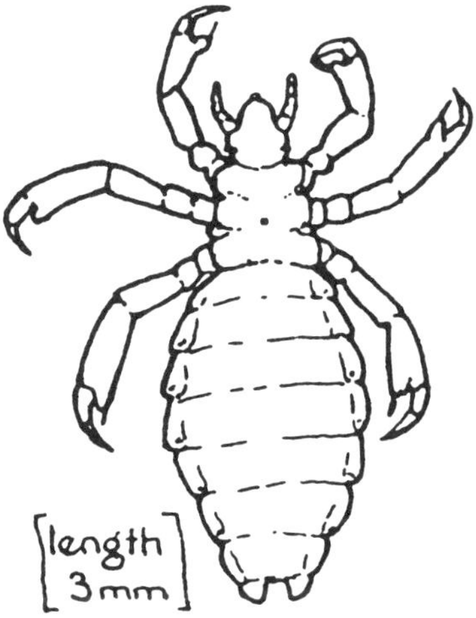
[length 3mm]
HUMAN HEAD/BODY LOUSE FORM

6.1

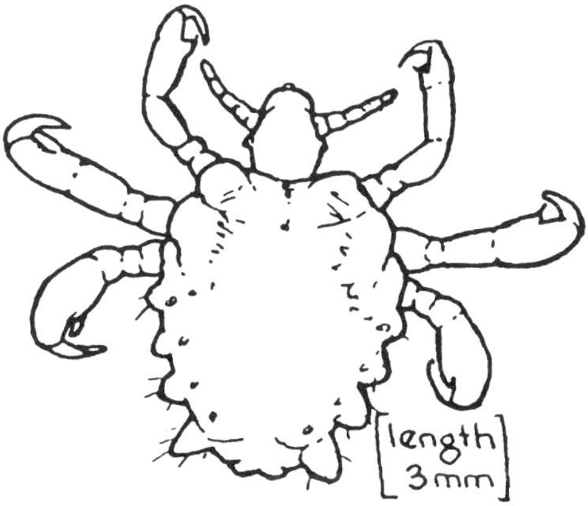
[length 3mm]
HUMAN PUBIC/CRAB LOUSE FORM

6.3

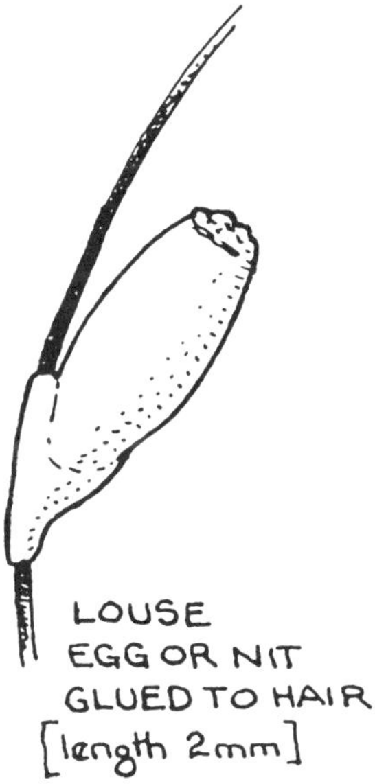
LOUSE EGG OR NIT GLUED TO HAIR
[length 2mm]

6.2

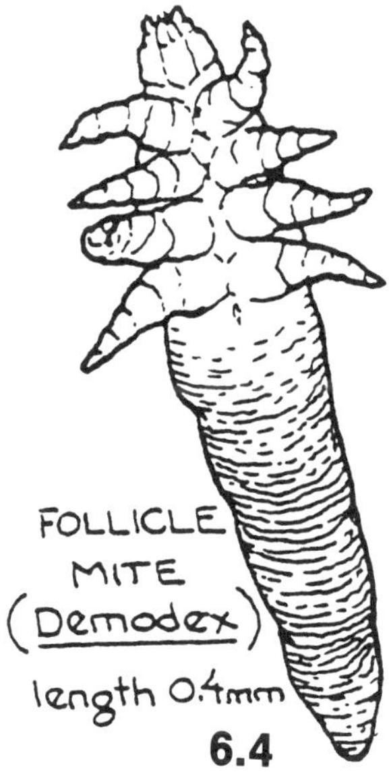
FOLLICLE MITE (Demodex)
length 0.4mm

6.4

Once the internal portion (e.g., surgical entry into the torso and skull) of the autopsy has begun, major sites of insect activity may include the skull with natural body openings, hair and scalp, respiratory tract (including inner nasal passages), esophagus, genital and rectal areas, and antemortem wound sites, as well as the chest cavity, pelvic area under desiccated skin, or areas under desiccated skin of the limbs. Insects collected from any of these body areas should be labeled, and preserved, and their location should be noted in the manner prescribed in Chapter 5.

COLLECTION OF SPECIMENS FROM BURIED REMAINS

When remains are buried, insect fauna is limited greatly by the amount of soil covering the body (Rodriguez and Bass 1985). Megnin (1984), Motter (1898) and Leclercq (1975) found that buried remains had their own insect fauna peculiar to a given type of environment. Burial slows the decompositional process due to lower, more constant temperatures at different soil depths and by exclusion of some bacteria and the majority of insects which colonize surface carrion. Blow flies can be excluded from access to buried remains by as little as one inch (2.5 cm) of covering (Rodriguez and Bass 1985). Sarcophagid (flesh flies) and muscid larvae have been recovered from remains which have been buried deeper. Several of the small flies from the groups Heleomyzidae and Phoridae as well as rove beetles (Staphylinidae) and ants (Payne 1968) have been collected from bodies covered with several inches of soil (10 cm) (Smith 1986).

Remains which are left exposed and unburied for a period of time, but buried later, may be colonized by species which are only capable of direct body access. The presence of such species may help establish a chain of events (a time period prior to burial, etc.) regarding the deposition of the corpse. For example, blow fly puparia recovered from 100 year old Indian skeletal remains from South Dakota suggested the season of the year in which the remains were buried (Gilbert and Bass 1967). A giant bison skull exposed by erosion from a stream bank in Alaska was collected for study. The internal cavity of the skull contained two species of fly larvae and puparia (**one species was identifiable**) which were carbon dated back approximately **22,000 years** (Catts, pers. comm.)

It is **essential to have a person experienced in proper exhumation techniques** and procedures present when unearthing buried remains. Damage can be caused to the remains when improperly exhumed. Excavating equipment (e.g., backhoe) **is not** recommended for this type of work, even though this is used by some police agencies! The damage caused may lead to the destruction of critical evidence with regard to cause of death or other injuries suffered by the victim before death (C. C. Snow, pers. comm.). A forensic anthropologist or archeologist is an excellent source of information and assistance for situations dealing with buried remains.

Collection of entomological specimens from burial sites is very similar to procedures described in Chapter 5 (Collection of Specimens After Body Removal). Collection of soil samples for later examination for insect parts or life stages should be undertaken by the entomological investigator as the excavation of the burial site proceeds. Soil from the ground surface down to the upper surface of the body should be sifted and examined as well as soil from the side and bottom of the burial pit after body removal. Insect larvae, puparia, adults or their parts may be found in the surrounding grave. If life stages are seen on the remains, collection of samples for preserving and rearing from the exposed insect populations should be made, but with care not to cause postmortem artifacts on the remains (see Chapter 5).

COLLECTION OF SPECIMENS FROM ENCLOSED STRUCTURES

Enclosed structures present several problems for the entomologist in the evaluation of insect colonization, development, and evidence collection. First, if the enclosed structure is very tightly closed (e.g., newer automobiles with the windows and doors closed; tightly sealed rooms; metal containers with air seals; newer, well insulated houses closed for air conditioner cooling), the chemical attractant cues for insects do not dissipate as rapidly or in a manner similar to bodies which are left in the open. The question arises as to how much time has elapsed before the odors finally emanate from the restrictive confines. In some circumstances, the odors never permeate from the container. However, even in such cases, extensive visual searches should be conducted at the death scene to make certain that no insect life stages are present.

Second, even if the attractant odor has emanated from the structure or container, insects may still be excluded from direct access to the corpse. Thus, there may be considerable numbers of flies or other carrion insects outside attempting to gain access to the corpse. The concentration of blow flies outside a structure or container (e.g., car trunk), most likely will indicate that something inside the enclosure is dead and beginning to decompose. Flying insects should be collected by aerial netting and sticky traps, and ground crawling insects should be hand collected by using collection methods described in Chapter 5. The duration of time that the remains have been in place may be indicated by certain assemblages of insects from a successional wave group (Megnin 1894). This is due to the production of different odor attractants or other chemical cues associated with advancing decomposition. For example, if only blow flies are found attempting to gain access, this may indicate a shorter PMI, whereas, the presence of phorid and muscid flies would suggest a more extended period of time.

Third, the enclosed structure probably will not have inside temperatures comparable to those reported from the local National Weather Service station. As explained in "Collecting Climatological and Temperature Data" (Chapter 5), temperatures in a particular outdoor habitat may differ considerably from those recorded at the NWS station, and independent temperature data must be collected from the scene to determine environmental conditions more accurately. It is critical to know the microclimatic environmental conditions of the location of the insects' development to estimate accurately the duration of their developmental time. This is even more critical when dealing with enclosed structures. A car parked on a black asphalt surface with the windows closed, even on a mild sunny day (75F or 24C), may exhibit temperatures 30 - 40F (20C+) degrees higher inside the passenger compartment or trunk. This condition could accelerate larval development by a degree equivalent to several days, thus creating a considerable error in PMI estimation. By following the procedures described in Chapter 5, or recreating the circumstances of the enclosed structure and recording temperature data, an accurate correlation may be obtained between the NWS data and the enclosed environment in question.

When investigating a death scene in a house closed tightly for heating or air conditioning, always check the thermostat setting controlling the heating or cooling system. Often, with the heat or air conditioning on, constant temperatures have prevailed for the period in question. This allows for a more accurate estimation of the rate of development of insects. Correlation of these data with on-site temperatures should be made even though this may be a repulsive task in a closed house containing an extensively decomposed body. The first person in the dwelling undoubtedly will open every window in an attempt to dispel the nauseous odors. In this event, the actual temperature to which the insects were subjected may be impossible to evaluate accurately until the building is closed again and the heating or cooling stabilized.

Collection of specimens from enclosed structures may entail knowledge of some of the likely places where the insects may migrate. In houses where the corpse is in advanced stages of decomposition (skeletonization and mumification), it is possible that more than one generation of larvae have migrated off the remains, changed to pupae, and emerged as adults. Inspection of the edges of the room where the walls and floor make contact, as well as, under the carpeting or carpet pad may reveal reddish to dark brown to black elongated fly puparia along the baseboard or in cracks or gaps in the wood. Often, post feeding maggots may be found under carpeting, rugs, or any other covering which may provide protection and seclusion for their transformation to the pupal stage. Crime scene investigators often ask why there were so many "rat droppings" in the room where the corpse was found. These are most likely not rat droppings, but are fly puparia. Close examination will show segmental lines (sutures) on the fly puparia; rat droppings are not segmented. Also, fly puparia will not dissolve in the mouth (the ultimate test).

The larvae may migrate to pupate into other rooms as well. Larvae may migrate up to 150 feet (50 m) in an open habitat, so maggot migration in an enclosed structure can extend to the outer limits of the structure. A basement or crawl space under this structure will be an additional source of insects if the corpse is in advanced decomposition and fluids have seeped through flooring and into the space below. Larvae, pupae, or adults may be found in such areas. When dealing with remains in automobiles, one should look under the floor mats and carpeting, down between the seats, and even under the upholstery which is in close proximity to the corpse. Larvae also may migrate into the car trunk to pupate, so this area, too, must be inspected thoroughly.

Because newly emerged adult flies and beetles will be seeking light and the outdoors, the back side of window blinds, shades and curtains should be searched for adult flying insects. Window sills and ledges on the inside of the structure may hold high numbers of adult flies or other species which have completed development on the remains, have emerged from puparia, and are seeking natural outdoor habitat. If the enclosed area is too hot, these insects may die from exposure to high temperatures. Thus, empty puparia, parasitized larvae or puparia, or dead larvae or adults may be found inside closed structures.

Dark fly specks (fecal spots) and lighter colored food regurgitation spots are deposited by many of the flies on the surrounding ceilings, floors, and walls or even on the corpse. Take note of the density of this spotting; it gives some indication of the relative size of the fly population attracted to the corpse. In advanced decomposition, insect feces (frass), can accumulate in conspicuous amounts around dried or mummified remains. Usually the cast skins of immature stages will be mixed in with the fecal material. In some cases, the dried feces still may be inside an extruded intestinal sheath, the peritrophic membrane. A mass of frass from dermestid beetles has the general appearance in color and form of pencil shavings (Fig. 6.5, also see chapter 2, case 19).

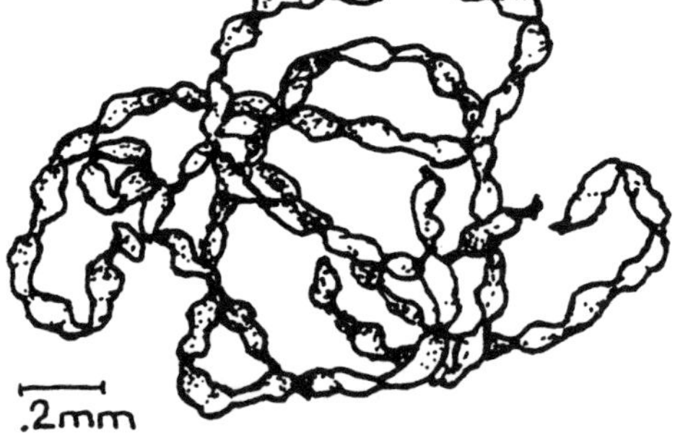

6.5 DERMESTID BEETLE FRASS (FECES) IN PERITROPHIC MEMBRANE

Basically, the same collection, preservation, and labeling techniques described in Chapter 5 should be followed when collecting from these enclosed environments. Dead dried specimens must be handled very carefully because they are extremely fragile. They should be stored in fluid perservative to allow for some rehydration.

COLLECTION OF SPECIMENS FROM AQUATIC HABITATS

At times it may be necessary to collect entomological evidence from bodies found in water. Sites may include shorelines along large bodies of water; brackish coastal backwaters; river banks; log jams or eddies in large rivers; rocks in shallow, fast flowing rivers; small, slow moving man-made lakes; still ponds; irrigation or drainage ditches; sewage ponds; sewers; open wells; or rain barrels. In each of these habitats, specific insect species may have special survival adaptations unique to these environments. Specialized insects may help the investigator identify a particular geographic location or a specific time of year.

Generally the aquatic insect species are not known to be necrophagous (specifically finding and then feeding on dead tissue) as are the terrestial species (Haskell et al. 1989). However, some aquatic insects will use submerged or floating corpses as a solid surface for attachment to facilitate feeding or shelter. Feeding usually will not include consumption of tissues from the body, although in some instances aquatic arthropods (e.g., crawfish, crabs, or shrimp) will feed extensively on human tissue (Smith 1986). Most often, aquatic insects attach to the corpse as a substrate for feeding on algal growth, finding secure places for attachment to filter-feed on small organisms, or hiding from predators.

Aquatic insects usually spend a major portion of their life cycle in water. Some species spend several years in water as a developing immature form before reaching the adult stage. In some cases, all life stages may be found in aquatic habitats. In many species, the growing and feeding stages are in water while the reproductive portion of the life cycle (winged flying adults) is usually of short duration and terrestrial for mating and dispersal. Therefore, death scene investigators usually will encounter only larvae, pupae, or other immature stages on submerged bodies, but may encounter newly emerging winged adults on corpses found on the surface or along shore lines or river banks.

While common terrestrial insects (e.g., blow flies, flesh flies, histerid beetles, carrion beetles) usually are not found on submerged bodies, floating bodies may contain many insects common to terrestrial scenes. These insects should be handled by using the same procedures described previously. If a corpse was colonized by terrestrial insect species prior to immersion, these insects may have drowned. This could indicate that the body was on the surface of the water for sometime before it sank to the bottom.

Some police agencies have procedures detailing recovery of bodies from water habitats. The protocol usually calls for some type of shroud or sheet to be fitted beneath the corpse before movement (if the corpse is floating) or for the body to be placed immediately upon a shroud if partially in the water. The shroud helps retain evidence on the body which may otherwise flush off when disturbed. This procedure is essential for proper recovery of aquatic insect specimens on the corpse. Most insects living on aquatic substrates will quickly detach at the slightest disturbance. Therefore, if a sheet or large, fine weave mesh net can be slipped under and around the corpse before it is disturbed, most of the insects using the body as shelter or as a food source will be captured in this shroud.

Proper collection and preservation of the aquatic insects is essential for accurate identification of the organisms. The specimens may be collected for either preserved specimens or collected as live specimens for rearing to adults. There are several standard techniques available for handling these fragile, soft bodied insects for rearing or preserving when investigating a death scene (Merrit and Cummins 1984).

Live Specimens

Live immature specimens should be collected and transported in water taken from the environment in which they are found. During transportation, the collecting jars should be filled completely to reduce damage to the insects caused by excessive splashing. Also, the water must not be allowed to elevate in temperature because many of these immature insects are heat sensitive. Keeping the collecting jars shaded, covered with a wet cloth, or against chemical ice will help reduce excessive heating during transit. Therefore, the collecting jars (any watertight small vial up to half-gallon size) containing the specimens could be packed in a styrofoam or similar type ice chest with ice or icepacks inside. Also, portions of naturally occurring substrate (identical species may be found here) found in close proximity to the corpse should be collected and placed into separate collecting jars. Appropriate rearing techniques are discussed in Chapter 7.

Preserved Specimens

Collecting procedures for preserving aquatic insects are very similar to those described under general collecting procedures (Chapter 5). The processing and labeling techniques also are the same. Data labels always should be placed in the vials immediately after the collection is made using the same label format as discussed previously (Appendix A.3).

Collection of flying adult insects which are in close proximity to the body should be accomplished with an aerial net. It may be possible to hand-collect some adults from the surrounding vegetation or directly from the body. In general, the wings are extremely important for identification of many of these aquatic species, and these insects may be quite fragile; they should be handled gently. The contents of the aerial insect net can then be funnelled into a vial containing 70 - 80% alcohol (preferably ETOH) or a similar preservative.

Eggs, larvae, and pupae also can be picked off the body (substrate), by using forceps, fingers, or a small scraping tool, as long as extreme care is used to avoid causing any additional postmortem artifacts. If a net or shroud is not used in recovery, specimens should be collected at the body recovery site directly from the corpse before it is moved (versus at the time of autopsy). Many of these aquatic insects will crawl off or drop from the disturbed body and may not be found even after the corpse is placed in a body bag.

Often, aquatic immature stages are difficult to see due to their small size and their inconspicuous coloration. In one case, aquatic midge larvae (flies in the family Chironomidae) were collected from a corpse discovered in a river and were mistaken at first as red carpet fibers (Hawley et al. 1989). Therefore, the investigator must look very carefully and closely at the outer portions of the skin and clothing. A hand lens may be necessary for these observations. These chironomid larvae also may be found under the slime and algae which coats the skin or clothing. The larvae may not be visible until this slime covering is scraped off. Caddisfly (Trichoptera) larvae or pupae may be found on and in clothing from bodies found in fast moving streams. Once the stages are collected, they should be preserved in one of the solutions listed at the end of Chapter 7.

There has been very little utilization of entomological evidence when dealing with death investigations in aquatic environments.

Few references report aquatic insects involved in death investigation cases (Holzer 1936; Hawley et al. 1989). The many kinds of insects living in these environments, as mentioned above, often are not recognized as insects and are hard to see. Also, many are difficult to identify to species, and there is little known of their behavior. This situation could be improved by involving entomologists knowledgable of aquatic insects when bodies are recovered from aquatic habitats. At present, although there exists considerable potential, there is very little baseline data identifying what species will be found on submerged corpses in aquatic habitats. The aquatic groups which may prove useful in yielding PMI and location information for death scene investigations include midges, caddisflies, mayflies, and some other small Diptera. If these insects can be identified and tabulated, data relating to their developmental time intervals might be used for making a PMI estimation.

Soft-bodied stages of aquatic insects should be preserved in fluids to prevent extreme distortion of the morphological characteristics resulting from drying. In contrast, several of the larger hard-bodied adult insects (dragonflies, dobsonflies and damselflies) can be killed by using a dry killing jar method after which they may be pinned and labeled.

Additional information on techniques of collecting and preserving the aquatic insects can be found in McCafferty (1981), Borror, Triplehorn, and Johnson (1989), Peterson (1967), Simpson and Bode (1980), and Merrit and Cummins (1984).

CHAPTER 7

PROCEDURES IN THE ENTOMOLOGY LABORATORY

Neal H. Haskell

Purdue University, West Lafayette, IN

After insect specimens have been received at the laboratory, they usually require some form of processing before analysis. To maintain the proper chain of custody record, documentation of all samples received should be done immediately upon receipt of any specimens in the lab. For those specimens collected directly by the entomologist during the scene investigation or at autopsy, notation of the disposition of those samples should be entered on the case report sheet. If live specimens are not processed immediately, they may be held temporarily (e.g., 1-3 days) at refrigerator temperature (ca. 38F or 4C) to minimize developmental changes.

Live specimens must be transferred to a suitable rearing medium and then held under relatively controlled conditions. Some, or perhaps all, of the preserved adult insects should be put on pins or cardpoints and labelled for microsopic study. Specimens which are still in KAAD or XAAD should be transferred to a more appropriate preservative solution. Finally, soil/litter samples must be processed by extraction and sorting of the included fauna.

REARING PROCEDURES FOR FLIES (DIPTERA) AND OTHER INSECTS

Generally, beef liver has been used as a blow fly larval (maggot) rearing medium in the laboratory. Other larval rearing media may be sheep or beef kidneys, lean raw beef, artificially formulated rearing media, or canned dog or cat food containing meat (not fish) and meat by-products with high protein content (check label). A more standardized artificial rearing medium consists of a gel containing meat/bone meal, powdered milk, and powdered eggs (Harris et al. 1985, see recipe at the end of this chapter). Maggot rearing pouches may be fabricated from aluminum foil by folding a 6 inch ×

7 inch (15 cm × 18 cm) piece of foil into thirds horizontally and then thirds vertically ending with a rectangular piece approximately 2 inch × 2.5 inch (5 cm × 7 cm). This rectangle is then unfolded and the corners crimped together forming an open-topped three dimensional rectangular pouch (Fig. 7.1). Plastic Ziploc® sandwich bags, perforated with multiple pin holes also may be used as a pouch-type rearing container. Plastic containers permit visual observation of the specimens as they transform through different stages of development. Both plastic and foil containers can be discarded after use.

The pouch containing 3 - 5 oz. (90 - 150g) of liver or other rearing medium is placed in a vented, 16 oz. (pint size) cardboard or plastic container (rearing cup) having approximately 1 inch (2.54 cm) of medium size vermiculite or sterile sand in the bottom of the container (Fig. 7.1). Ten to fifteen live larvae (maggots) are introduced into the foil pouch. A wet piece of paper toweling is then placed over the liver, with a dry piece of toweling covering the wet piece. This will help prevent desiccation of the liver as well as the feeding larvae. The top edges of the aluminum foil pouch are lightly crimped (folded over) to lessen desiccation. This light crimping will allow the mature larvae to exit the pouch and pupate in the vermiculite. The cardboard rearing cup with the maggot pouch inside is held at room temperature or placed into an incubator with temperatures between 68 - 80F (20 - 27C) and at 60 - 75% relative humidity. If plastic bags are used, they should be filled one third full with wood shavings first and then with meat or an alternative rearing diet and the live larvae. A small piece of "starter" meat (a small cube of beef liver) is put with the artificial diet and will be consumed in the first few days of maggot development.

Upon emergence, adult flies, immobilized by cooling (temperatures at or just below freezing), are either pinned or placed in 75% ethyl alcohol, labeled, and stored for later identification. It is necessary to record actual temperatures during the development period in the laboratory. Also, the date and time of adult emergence (eclosion) should be noted in the case record notes and placed on the date/location label. If necessary, eclosed adults can be maintained alive in small cups with dental wicks saturated with sugar or similar liquid (e.g., Gatorade® or Koolaid®). When labeling pinned reared specimens, the date and, if possible, the time of emergence is included on the label. With this system, when two dates are seen, it is possible to recognize immediately that the specimen has been

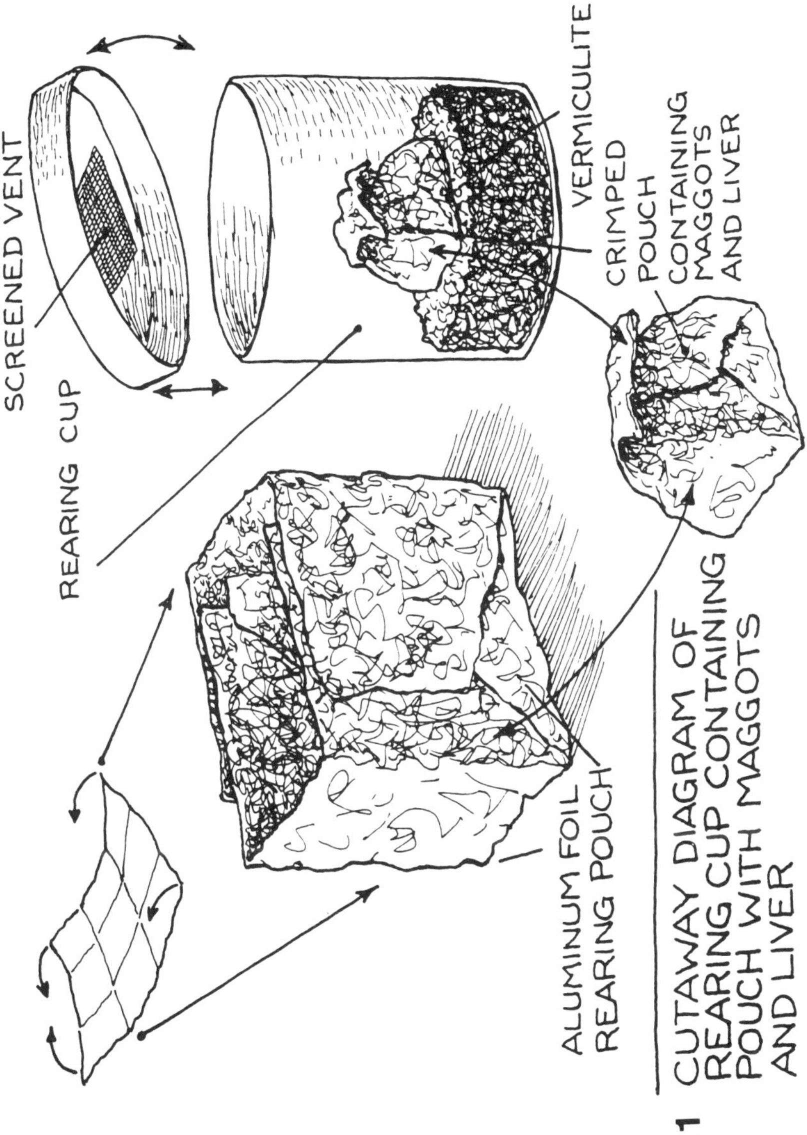

7.1 CUTAWAY DIAGRAM OF REARING CUP CONTAINING POUCH WITH MAGGOTS AND LIVER reared, and the first date being the date collected and the second date being the date emerged. This system also is used when rearing eggs or larvae to different stages (see Appendix A.3).

Eggs are treated in the same manner as the maggots. Live puparia can be placed directly into the rearing cups which have vermiculite or sand in the bottom or in the plastic bags containing woodchips only. Puparia are lightly covered with vermiculite. It is not necessary to provide the pouches with rearing medium because this stage does not feed.

At the death scene, beetle larvae [generally larvae with three pairs of legs (see Chapter 3)] also may be collected along with the maggots. Many beetle larvae are predaceous on maggots. Thus, these live beetle larvae should not be placed in the same container with the live maggots, or only beetles will survive. Live beetle larvae need to be placed into separate containers for rearing. If the beetle larvae are recognized as being predaceous on maggots, extra maggots should be collected for use as a food source for them. These feed-maggots may be held over an extended period under refrigeration (38F or 4C). Otherwise, portions of rearing medium may be used as a food source for immature beetles.

Most adult insects will be in a winged form and will need to be restricted from escaping once they emerge. Adult flies or beetles emerging in the rearing cups or plastic pouches will be confined, but caution should be used when inspecting the cups when close to the time of eclosion (emergence) to prevent specimens from inadvertently escaping when the containers are opened. Adult flies usually eclose in the morning and require a few hours exposure before they are capable of flight. Flies should be maintained alive for about 48 hours before killing and preserving them. This will assure full pigmentation and wing formation of the specimens.

REARING OF AQUATIC INSECTS

Aquatic insects can be reared fairly easily, except for those that require high velocity stream flow or riffle simulation. A small or medium size aquarium will work adequately for rearing of most immature aquatic insects. If possible, the insects should be kept in water from the collection site. If tap water is used, it should be dechlorinated by letting it stand overnight. Rain water collected in buckets or other containers and then placed into the tank or aquarium is also suitable. Distilled water can be used to replace that lost to evaporation.

Placement of some of the substrate found at the site in the rearing tank should be attempted along with objects for the insects to crawl or float on. Emergent sticks, rocks, leaves, grass or other material found at the site can be used for this purpose. Usually aeration of the tank is needed unless samples are collected from stagnant water. Aeration can be accomplished by placing a small aerator stone or water pump in the rearing tank and directing the water flow in a manner to create oxygenation of the water. Many aquatic insects are feeders on algae or detritus, others are scavengers, and some are predatory. Feeding needs for most of the insects will be met by submerging decaying plant material found at the scene along with substrate found there. This can be supplemented with commercial fish food. The predacious species need to be provided with animal forms comparable to their prey and must be reared in separated containers.

If a small tent type apparatus is constructed on top of the rearing tank, emerging aquatic adults will be captured after they leave the water to take flight. If the tent is made in a pyramid type configuration with a small closable flap or access door at the very apex, the adults can be removed by means of this access door. Then adults can be handled in the manner prescribed for preserving adult flies.

Aquatic insect rearing containers are commercially available for rearing small insects such as mosquitos, midges or other small flies (Diptera). These containers are of a two-chamber construction with the lower chamber for growth and development of the aquatic immatures and the upper chamber for entrapment and collection of the newly emerged flying adults (Fig. 7.2). Such containers also are suitable for rearing terrestrial flies, especially when puparia are placed into vermiculite in the lower growth chamber and allowed to eclose.

PROCESSING LITTER AND SOIL SAMPLES

The insects and other arthropods (especially mites) which live in litter and humus can be collected by using a Berlese/Tullgren funnel (Fig. 7.3). The funnel contains a wire screen platform to hold the litter in place but allows insects and mites to pass through the screen and down through the funnel. A small bottle or vial containing

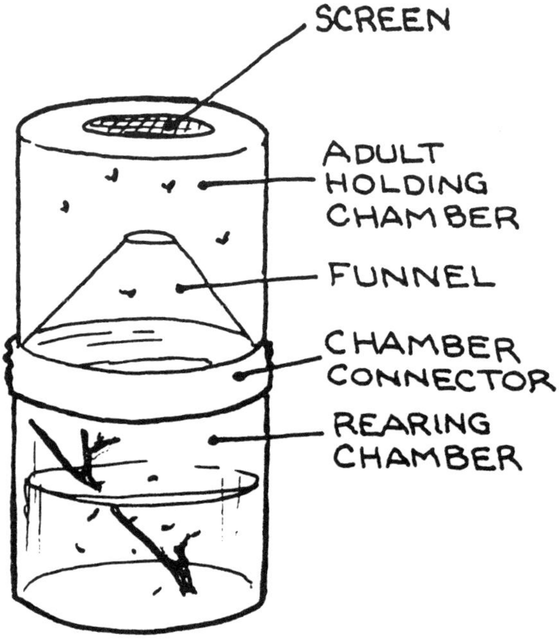

7.2

REARING CHAMBERS FOR AQUATIC INSECTS

75% alcohol at the bottom of the funnel cone captures the tiny animals. Insects and mites are driven to the bottom of the funnel by heat from a low wattage light (10 - 25 watt) or by fumes from a chemical repellant. Litter samples are placed in the funnel on top of the wire screen. The light or cloth bag containing the chemical is suspended a few inches over the sample material and a vial or jar of alcohol is placed under the lower funnel opening. In three or four days most of the insects and mites will be driven from the litter material and into the alcohol bottle. These specimens can then be processed and examined.

Soil samples can be examined by using different size meshes of metal screens. First, a large ¼ inch (6 mm) opening mesh screen should be used to separate large particles of soil and large insects. Sifting the screenings onto a large flat white pan helps keep all of the evidence confined and allows the material to be spread out in a thin layer to expose active insects. The white background also makes the smaller, moving insects more visible. Again, the insects

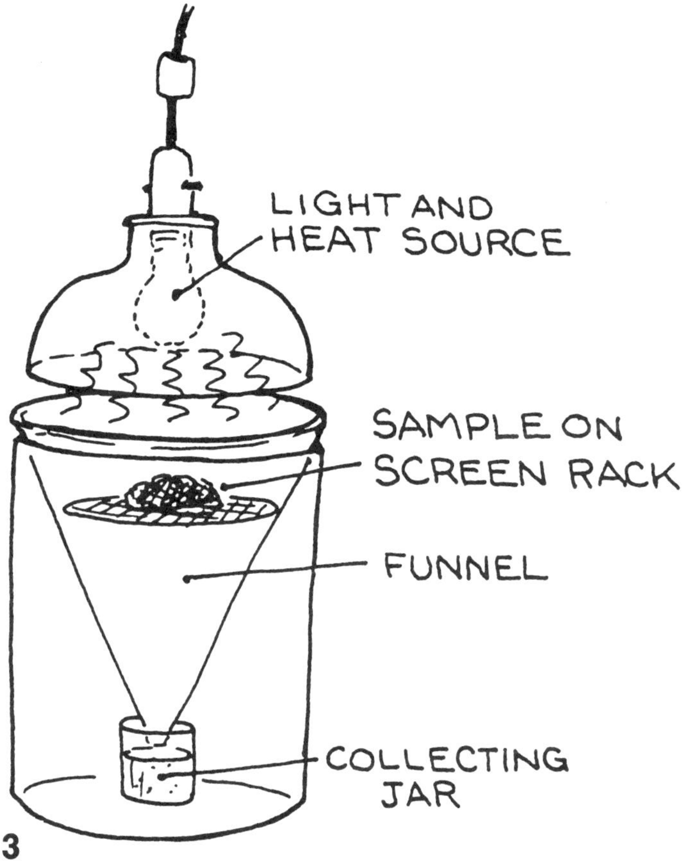

7.3

BERLESE/TULLGREN FUNNEL

should be preserved in the same manner as described earlier: adults in 75% alcohol and larvae in Kahle's or 75% alcohol. After these screenings are examined thoroughly, the finer particles are screened again using a 1.4 mm mesh screen. This will separate the large larvae, pupae, and the medium size adults from the very tiny adults and larvae. The white background of the pan will contrast with any of the very small insects, especially if there is any movement, thus enabling those to be seen and collected. A technique used for picking up very small insects and mites is to dip one point of a forceps or a fine artist's paint brush into the collecting vial of preservative solution and touching the wet tool to the insect (Fig. 7.4). The surface tension of the liquid will hold the insect as it is transferred to the collecting vial and the insect is "washed off" by the solution. Another method is to use a blow-type or electric aspirator to suck up the insects into a collecting chamber (Fig. 7.5) for placement into ethyl alcohol. Mites may be collected from the soil samples by using the Berlese/Tullgren funnel technique for the soil sifted through the 1.4 mm screen.

Insect movement and mobility can be reduced if the boxes containing the soil samples are placed in a refrigerator at 34 - 40F (1-4C) for approximately an hour prior to sorting or screening. Notes should be made of the time duration the samples were in the refrigerator and of the associated temperatures if the time exceeds one hour because prolonged cold temperatures will retard development of immature insect stages.

The entomologist also should keep in mind that as the leaf litter or soil samples are being processed, other physical evidence may be included. Bone fragments, hair, teeth, projectiles or other physical evidence essential to the case may be contained in samples which the entomologist examines. This evidence should be handled in the manner prescribed for collecting and labeling any other physical evidence (Chapters 4 and 5) and should be reported to the primary investigator and/or forensic pathologist.

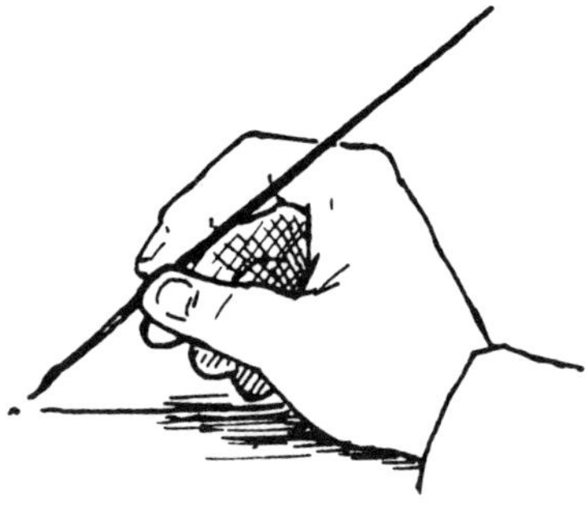

7.4

PICKING UP SMALL CRITTERS WITH A WETTED BRUSH

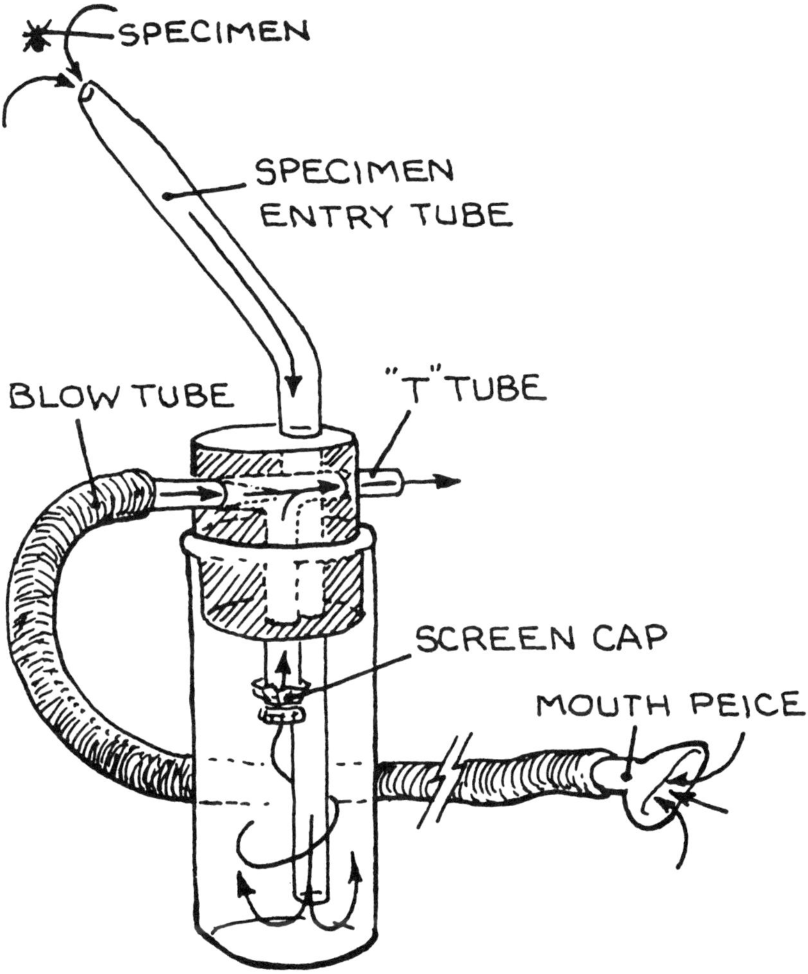

7.5

BLOW-TYPE ASPIRATOR
(CUTAWAY DIAGRAM)

SOLUTIONS FOR KILLING AND PRESERVING INSECTS

1. Ethyl Alcohol (Ethanol) 70 - 80% solution

   For killing and preserving adult specimens; for preserving larval specimens.

   Ethanol (ETOH) usually is purchased in bulk at 95% concentration. To make a 70% solution from this, simply measure 70 units of the 95% ETOH and add enough water (distilled is preferred) to give a total of 95 units of mixed solution. An 80% solution is similarly prepared by adding enough water (15 units) to 80 units of 95% ETOH to give a total of 95 units of mix.

2. Hood's Solution:

75% Ethanol	95 ml
Glycerine	5 ml

   For killing and preserving adult specimens; for preserving larval specimens.

3. Kahle's Solution

95% Ethanol	30 ml
Formaldehyde	12 ml
Glacial Acetic Acid	4 ml
Water	60 ml

   For killing and preserving adult specimens; for preserving larval specimens.

4. KAAD:

95% Ethnol	80 - 100 ml
Glacial Acetic Acid	20 ml
Kerosene	10 ml
Dioxane	10 ml

   For killing larval specimens only (not for preserving as it causes specimens to become overly brittle after approximately 12 hours.) After killing, specimens should be transferred from KAAD to 70 - 80% alcohol (ethanol).

5. X.A.:

Xylene	50 ml
95% Ethanol	50 ml

For killing larval specimens.

6. X.A.A.D.:

Xylene	40 ml
Isopropyl Alcohol	60 ml
Glacial Acetic Acid	50 ml
Dioxane	40 ml

For killing larval specimens only (not for preserving). See comments for K.A.A.D.

The solutions intended only for killing larval specimens are best used at the point of initial collection. The larvae should then be transferred into 75% ethanol within 10 - 12 hours after they have been placed in these killing agents. The longer they are left in the killing solutions (i.e., KAAD and XAAD), the more brittle they will become. If these are shipped to the entomology laboratory in the killing agents in less than a day after collection, then the entomologist can make the tranfer to ethanol.

PRESERVATIVE SOLUTIONS FOR AQUATIC INSECTS

1. 90 to 95% Ethyl alcohol..........................eggs larvae, and pupae

2. Carnoy fluid ....Soft bodied insects

Glacial acetic acid	10 ml
Ethyl alcohol 95%	60 ml
Chloroform 30%	30 ml

3. Hood's solution ....all stages
   see p. 120

Strong penetrating solutions for use when large amounts of extraneous material accompany the samples.

4. Kahle's solution...all stages
   see p. 120

5. Pampel's solution...all stages

Formalin (10% Formaldehyde in water)	10 ml.
Ethyl alcohol 95%	30 ml
Glacial acetic acid	7 ml.
Water	53 ml.

6. 75 to 80% Ethyl alcohol....................................Soft-bodies adults

ARTIFICIAL FLY REARING DIET RECIPE (KERRVILLE BLOW FLY DIET — Harris et al. 1985)

(14:3:3 ratio by weight)

Ingredient weight	**Ingredient**
140 g	Meat and Bone Meal
30 g	Powdered Milk
30 g	Powdered Egg
5 g	Carrageenan (No. C-1013 — Sigma Chemical Company, St. Louis, Missouri) — a gelling agent
1 ml	Formaldehyde (to reduce mold)
597 g	Hot tap water (**65° C**)

Supplement with a whole meat starter (e.g., a small cube of beef liver).

SELECTED REFERENCES FOR COLLECTING AND PRESERVING INSECTS

Borror, D. J., C. A. Triplehorn, and N. F. Johnson. 1989. An introduction to the study of insects, 6th Edn. Sanders College Publishing, Philadelphia, 875 p.

Furman, D. P., and E. P. Catts. 1982. Manual of medical entomology, 4th Edn., Cambridge University Press, Cambridge, 207 p.

Hall, D. G., 1947. The blowflies of North America. The Thomas Say Foundation, Baltimore, 1947, 477 p.

Harris, R. L., E. F. Gersabect, E. Corso, and O. H. Graham, 1985. Screwworm larval production on gelled media. Southwest. Entomol. 10: 253 - 256.

Lord, W. D. and J. F. Burger. 1983. Collection and preservation of forensically important entomological materials. J. Forensic Sci 28: 936-44.

McCafferty, W. P. 1981. Aquatic entomology. Jones and Bartlett Publishers, Inc. Boston, 448 p.

Nuorteva, P., M. Isokoski and K. Laiho. 1967. Studies on the possibilities of using blowflies (Dipt.) as medicolegal indicators in Finland. 1. Report of four indoor cases from the city of Helsinki. Ann. Entomol. Fenn. 33: 217-25.

Peterson, A. 1967. Larvae of insects, and introduction to nearctic species, 6th Edn., Printed for the author by Edwards Brothers, Ann. Arbor, Michigan, 312 p.

Simpson, K. W. and R. W. Bode. 1980. Common larvae of the Chironomidae (Diptera) from New York state streams and rivers, with particular reference to the fauna of artificial substrates. Bull. New York St. Mus., Bul. No. 439, 1-105.

Smith, K. G. V., 1986. A Manual of Forensic Entomology. Cornell University Press. Ithaca, New York. 205 p.

CHAPTER 8

ANALYZING ENTOMOLOGICAL DATA

E. P. Catts

Washington State University, Pullman, WA

There are five questions posed during a death scene investigation: **who** is the deceased, **how** did death occur, **where** did it occur, **when** did it occur, and is this a **natural,** an **accidental** or a **criminal death**? Forensic specialists working within different disciplines help answer these questions. The forensic entomologist's role is primarily to determine when death occurred. This information is a major element in any death investigation. Even in a case of suicide where a dated note indicates the time of death, faunal evidence should corroborate the alleged time in the note. If it does not, then the validity of the note must be questioned.

Entomological evidence also can be involved in answering the question of where death occurred. Certain insects and mites, that take a period of time to develop, become established in distinctive subcorpse communities. The absence of this community indicates either that the body has been moved, suggesting that death occurred elsewhere, or that the remains were inaccessible to insects.

Insect evidence also may help in determining the manner of death. Sites of concentrated insect activity in a corpse often reflect locations of antemortem injury. In child and elderly neglectful and physical abuse cases, the kinds of insects present in the corpse may indicate antemortem insect-invasions and thus indirectly suggest antemortem injury.

This chapter is intended only as a guide for analyzing entomological data in death scene investigations. Each case has its own set of circumstances and peculiarities that defy exact matching of events to a predetermined scenario. However, there are some basic assumptions and principles that should be kept in mind.

WHY USE INSECTS IN DEATH INVESTIGATIONS?

There are several reasons for using insects and other arthropods as evidence in death investigations.

1. Insects are generally the first to discover a corpse, particularly if any attempt has been made to hide the body. Often in outdoor death cases, blow flies will arrive and begin laying eggs on a body within several minutes of death.
2. Insects arrive at a decomposing corpse in a somewhat predictable sequence. This is termed ecological succession and consists of a series of blending waves (seres) of different arthropod actors. Differences in the cast of actors can be used to describe incidents surrounding the death.
3. Often insects are ignored or even discarded as evidence at a death scene or an autopsy. Our culture generally views maggots and flies as repulsive intruders to be avoided or destroyed. Unfortunately, this attitude is sometimes held by death scene investigators and coroners. Conversely, the absence of insects in situations where they should be present is valuable evidence in any investigation; the absence of insects also demands inquiry.

ESTIMATING THE PMI

The major contribution made by forensic entomologists in death investigation is an estimate of the duration of the postmortem interval (PMI). An accurate PMI is of primary importance in any homicide case because it narrows the field of suspects and may aid in identification of the deceased with reference to known missing persons. The entomologist's estimate is based on a series of generally valid assumptions, but variance, or error, in any one of these assumptions can skew the accuracy of the estimated PMI.

Estimating the PMI involves the setting of a minimal and maximal probable time interval between death and corpse discovery (Fig. 8.1). The maximal limit is determined from the species of insects which are present and the weather "windows" available for activity of these species. The known habits and mix of these species can be used to make a preliminary estimate of the earliest time of corpse exposure to insect activity. Weather windows are intervals of conditions, usually warm temperatures and sunshine, when carrion seeking species are more likely to be active. The minimal PMI is determined largely by estimating the age of developing immature insects collected at the time of corpse discovery. The estimated time needed for them to develop without interference under the conditions present sets the miminal PMI.

Usually the maximal PMI is more difficult to establish. Clearly, it can be established with unwavering certainty as the time when the deceased was last seen alive. Beyond this, the maximal PMI is estimated by considering the extent to which other factors may have prolonged insect development. Such things as cold or inclement weather, disturbance of the remains by large scavengers, or maggot mass depletion by ants or other predators can have a disruptive effect on insect development and community composition. If these kinds of disturbances are negligible, then the maximal and minimal PMI estimates should be either close together or identical. For example, if the estimated minimal PMI corresponds to the time since the deceased was known to be alive (the maximal PMI), then the two can be considered identical.

Usually PMI estimations are based primarily on the early seral wave arthropods (e.g., blow flies or flesh flies). This requires careful interpolation of a few baseline laboratory studies which were done under a controlled laboratory regimen.

These studies rarely duplicate the conditions associated with any given death scene. Outdoor studies with local species and environmental conditions can constitute a better reference. This procedure then can account for possible local insect population differences in biological traits within a species.

BASIC ASSUMPTIONS

There are several assumptions that generally hold true in most homicide investigations:

1. Most homicides occur at night, **under cover of darkness when flies are presumably inactive.** The first part is true, but recently Greenberg (1990) demonstrated that blow flies may oviposit on small animal carcasses during dark of night. So the assumption that flies are inactive at night should be applied with caution. For most cases this is probably valid, but if the corpse is put close to a gravid female fly, darkness may not deter egg laying.
2. **Flies will begin ovipositing as soon as they discover the body.** This also is generally true, but the number of flies

actually ovipositing may be quite small in comparison to the total number of flies present. Thus there can be some delay before the bulk of the maggot mass begins development. This is why it is important to collect the largest maggots observed as part of the evidence sample even though they may be comparatively few in number.

3. **Faunal succession in and under the corpse will follow a predictable pattern.** Again, this is generally true, but the entomologist should recognize that at times certain species may appear out of sequence (wave jumpers) and should try to explain why this happened in a specific case (Bornemissza 1957).
4. **Weather conditions recorded at a site distant from the death scene reflect the conditions at the scene.** This is a weak assumption. Although gross weather changes may be quite similar, microhabitat conditions which have greater impact on insect activity and development, are rarely comparable. The averaging of daily temperature highs and lows also may give only a crude idea of actual conditons. The low temperature may have lasted for only a few hours or less, and yet it is given equal weight with the high. Consequently calculations of accumulated degree-hours (ADH) based on mean temperature values to estimate the duration of maggot development can be misleading. Indoor temperatures are much less variable and ADH calculations have more validity in that scenario. Outdoors or indoors, a remedy is to record temperature and humidity at the death scene for several days following corpse removal to validate reasonable interpolation of consistent differences with those of the more distant recording station.
5. **Ambient air temperatures are the major factors influencing the rate of maggot development.** Ambient heat does play a dominant role during egg and early larval development, but after that its effect decreases rapidly. Maggot activity generates its own heat that can prevent cooling and slowed development when air temperatures drop. The early maggot mass will develop at a rate somewhat in synchrony with fluctuations in ambient temperatures. However, in the late 2nd and 3rd instars, heat generated by the massed maggots can far exceed ambient heat and in effect produces a stable optimal heat level for accelerated larval development. The warmth produced by the maggot-mass serves also to warm

the substrate underlying the remains so that subcorpse temperatures also may exceed that of the air. As the maggots from the mass scatter during postfeeding dispersal, the ground and corpse warming also dissipates.

Another important consequence of maggot-mass heating is that maggot development may not be arrested in a corpse held under refrigeration prior to autopsy. This means that if the maggot-mass is of reasonable size, the possibility that larval growth continued under refrigeration must be considered. Thus the temperature of the mass should be taken on autopsy. Ideally, at the death scene, temperatures of the maggot mass, the body surface and the substrate beneath the corpse also should be recorded (see Chapters 5 and 6).

ANALYSIS OF ENTOMOLOGICAL DATA

1. First, determine the approximate phase of decomposition of the remains (i.e., fresh bloat, active decay, advanced decay, putrid or dried remains — Fig. 8.1). Keep in mind that these phases are only a crude characterization of decomposition, and that different parts of the corpse may be in different phases of decomposition at the same time. From local experience and/or the literature, consider what arthropod groups and stages should be expected to be present given the phase of decomposition and the season. Next examine the specimens collected as evidence to determine if any anticipated groups are missing. If they are, arrange for additional collections to be made.

   If the corpse is fresh, verify with the medical examiner/coroner the presence or absence of skin parasites such as follicle mites, ticks, sucking lice or louse eggs (Figs. 6.1, 6.2). Although they are not part of the decomposer community, their longevity may help to pinpoint the time of death (Smith 1986).
2. Now identify (to family, genus and species if possible) the specimens collected in evidence and those reared in the laboratory. From the literature and your experience, determine the geographic and seasonal distribution of species and the stages of development and ages of the largest specimens collected. Record these identifications (Appendix A.4).

3. For outdoor cases, interpolate the postmortem environmental conditions from the weather record, on-scene measurements and consideration of other probable influencing factors such as slope air drainage, solar radiation exposure, burial, immersion in water, and vegetative or other cover. For indoor cases, determine the heating or cooling regimen for the approximated postmortem period from thermostat settings and degree of isolation from outside weather.
4. From the Autopsy Report or by direct observation, note the sites on the corpse of major faunal infestation. Also obtain the determined manner and cause of death from the medical examiner/coroner or from the Autopsy Report.
5. Determine from the death scene reports, if possible, whether or not the remains have been disturbed or disarticulated during the PMI. Floods and landslides can cause body movement. This also may occur where there are large scavengers about, and such movement will prolong fauna succession of the remains. If the subcorpse fauna does not correspond to the approximate phase of corpse decomposition, then consider that the remains might have been moved during the PMI.
6. Ask if there is any indication of the presence of antemortem administered toxicants or drugs including alcohol. Insects, such as maggots, which have ingested tissue from poisoned or drugged victims may be analyzed and may show traces of the substance used (Gunatilake and Goff 1989). Certain substances also may affect the rate of maggot growth (Goff et al. 1989).
7. Now taking the previous facts under consideration, estimate the age of as many insect specimens as necessary to make a reasonable and defendable approximation of the interval from the time of preservation of the specimens back to the probable time of the initiation of insect activity. Furthermore, consider whether or not insect activity was immediate or delayed regarding time of death (e.g., weather windows of opportunity, proximity to resting fly harborages, probable seasonal population levels of blow flies, accessibility of the remains to insects). Thus the final estimated minimal PMI often will be longer than the estimated faunal ages.

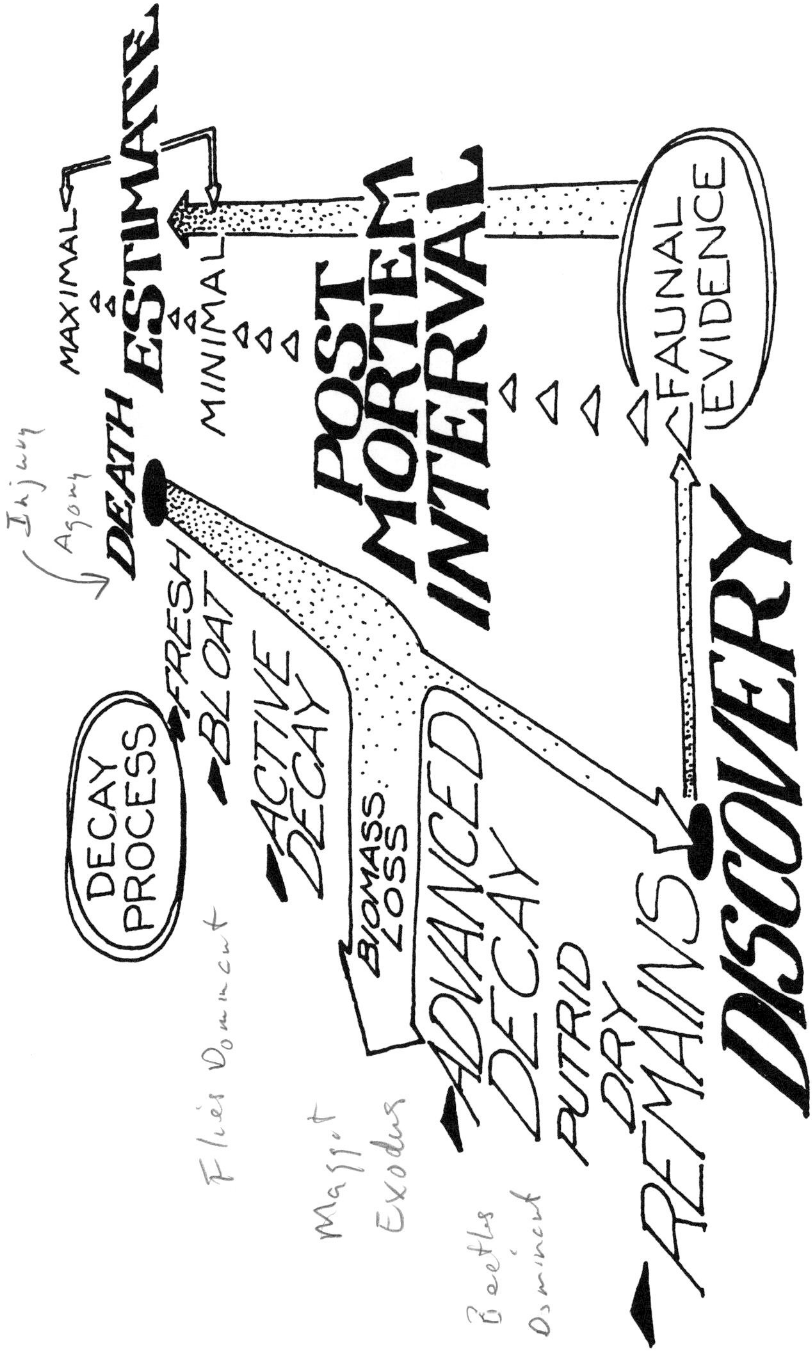

8.1 DIAGRAM OF CORPSE DECOMPOSITION TIMELINE

AGE ESTIMATION

An accurate estimate of the age of preserved maggots and puparia sampled at the death scene or on autopsy is key to making a credible and defendable estimate of the PMI. This involves the determination of the duration of development for eggs, larvae and pupae. There are several methods employed to do this. Each has some room for error. Basic input data for determining the age of maggots is the temperature of the corpse throughout the period of development and the size (length or dry weight) of the larvae at the time of collection (Fig. 8.3). Assuming that the eventual maggot mass was of sufficient size to nullify the effect of low ambient temperatures, the critical element becomes the ambient and corpse temperatures prior to egg hatch and during early maggot development. The methods used for maggot age determination generally involve a comparison of sample maggots to maggots of known ages which were reared usually under constant conditions (Kamal 1958; Nuorteva 1977; Smith 1986; Williams 1984). The developmental time estimate is then refined by interpolation to take into account the particular case circumstances (e.g., temperature extremes, habitat, shading, cover, etc.).

The estimate is sometimes expressed as Accumulative Degree Hours (ADH) or Accumulated Degree Days (ADD) and refers to the duration of temperatures above and below the lower and upper lethal maggot developmental thresholds, respectively. Unless corpse temperatures are known precisely, ADH estimates are specious and should be avoided because they tend to give a false perception of scientific accuracy that can be misleading.

In making a faunal age estimate, the entomologist must be able to recognize the morphological and physiological features that mark development times in the species examined. In flies and beetles, for example, the egg, larval and pupal forms (i.e., stages and instars) give a somewhat punctuated record of the insect's age. The predominance of 3rd instar maggots (L^{III}) and the knowledge of how long it takes to reach this point in development under the given conditions clearly indicates the age of these insects. Even finer age estimates can be made by dissecting specimens within a given stage or instar.

Some examples of this kind of refinement in age estimates follows:

1. Although fly eggs (Fig. 8.2) develop fairly rapidly, their development can be traced by the gradual appearance and delineation of the first instar (L^I) maggot within the egg. This will support an age estimate that corresponds to the point of embryonic development before hatching.

   The rate of body cooling (algor mortis) following death gives some indication of the time of death. In general internal corpse temperatures are taken at autopsy only when bodies are in fresh condition because the corpse will cool to ambient levels in 18 - 24 hours (Geberth 1990). The dissipation of inherent heat can be assumed and is important to fly egg development during early decomposition.
2. The internal mouthparts (cephalopharyngeal skeleton) and posterior spiracles of each maggot instar do not increase in size as the maggot grows during the instar. Thus the relative size of the mouthparts compared to the total length of the maggot indicates an age within the instar time period (Fig. 8.3). For these comparisons the larvae should be killed and fixed in very hot water. Heat from hot water will quickly coagulate internal tissues and the maggot will remain in a fully expanded posture.

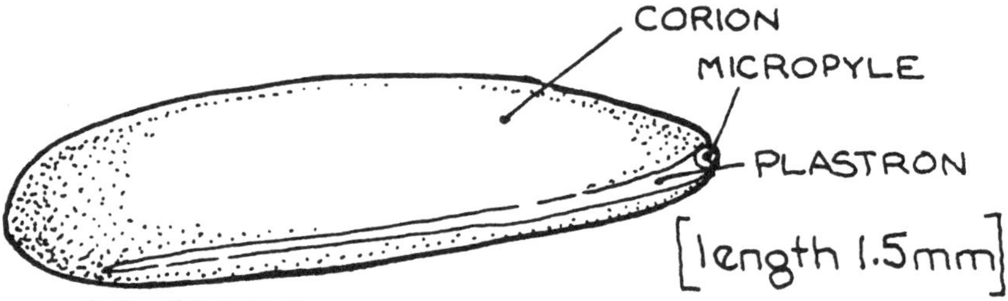

8.2 TYPICAL BLOW FLY EGG
(Eucalliphora sp.)

3. Just prior to molting, the spiracular structures of the subsequent maggot instar will be visible beneath those of the present instar. This indicates a very specific time in the life of the maggot.
4. Generally the darkened crop of an actively feeding maggot is visible through the integument (Fig. 8.3). However, when the

postfeeding maggot begins to wander in search of a pupariation site, the crop becomes greatly reduced in size. In addition, an opaque layer of fat body tissue is laid down beneath the maggot's integument. This layer will obscure all internal structures (Fig. 8.3) including cephalic structures. The degree of masking of internal structures by fat body tissue is another indicator of postfeeding maggot age.

5. Puparia not only change in external coloration as they age, but also undergo a predictable sequence of internal changes as the adult slowly develops and gains its own pigmentation. Pupariation and pupation make up about 40% of total development time for a fly, so knowledge of the sequence and timing of these changes will allow further refinement of the age estimate.
6. In more advanced corpse decomposition the complexity of corpse and subcorpse communities increases. This too can be used as an indicator of time of exposure to faunal exploitation. The assemblage of different corpse frequenting species and their various stages of development can suggest a fairly narrow time interval. This requires the identification of all species collected and an analysis of the character of their diversity. This kind of analysis by using a computer program "for rapid comparisons of total assemblages" has been employed successfully in several cases where the remains were in a more advanced state of decomposition (Goff and Odum 1987).

OTHER FACTORS

In cold temperate and arctic areas, very cold winter conditions may freeze and preserve the corpse for a prolonged period of several months. For example, in the winter in Montana a dead body outdoors can freeze solid and will not thaw until spring warming. Decomposition is arrested during the frozen state, but on thawing it will progress very rapidly. Solid freezing also will kill associated fauna. At high elevations where temperature extremes often are unpredictable and frequent, repeated freezing and thawing in a narrow time interval will greatly accelerate decomposition. Some insects can survive short periods of freezing to continue their development during periods of warming.

Animals other than insects or people also can affect corpse decomposition. Large scavengers (e.g., alligator, bear, mountain

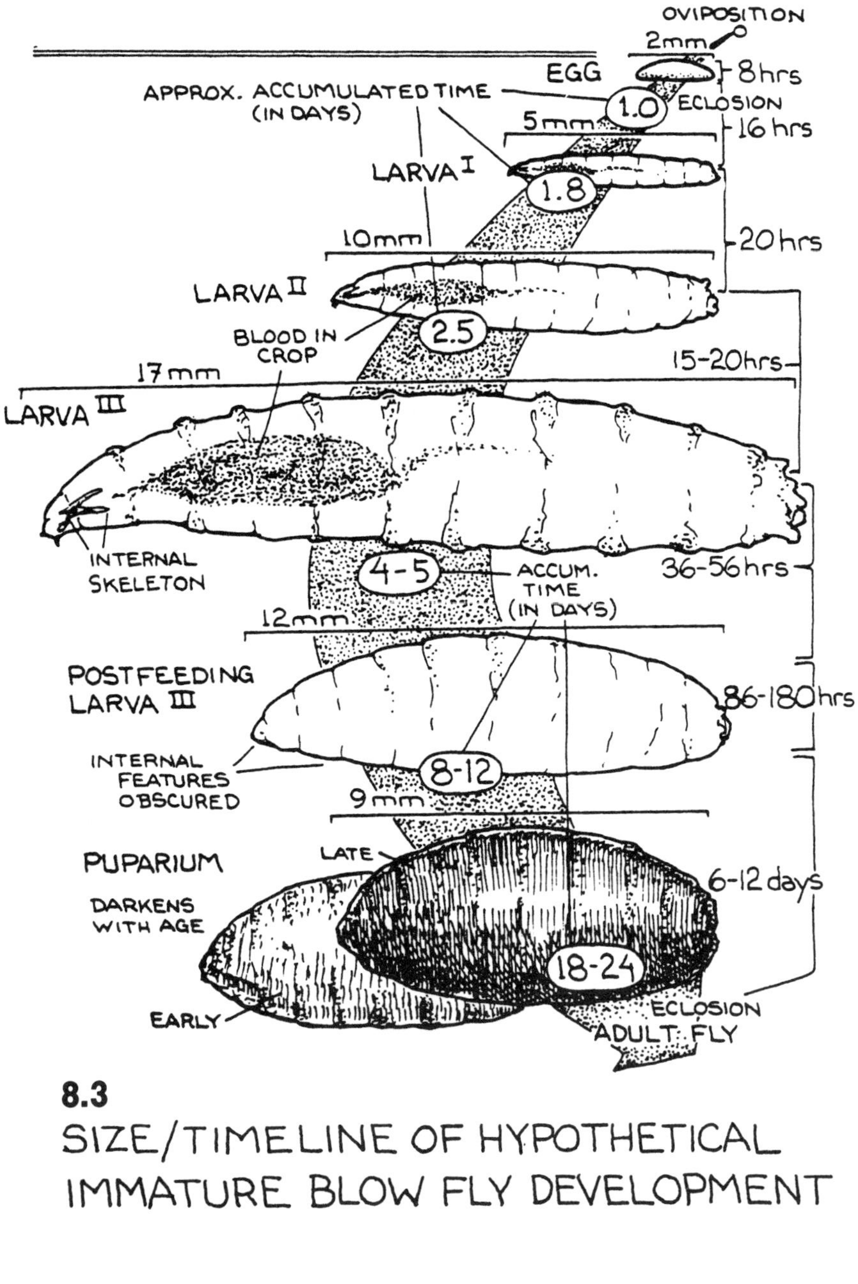

8.3

SIZE/TIMELINE OF HYPOTHETICAL IMMATURE BLOW FLY DEVELOPMENT

lion, coyote, feral dog, and vulture) may feed on and, in so doing, disturb the corpse and its fauna. With the exception of vultures, these animals generally are nocturnal and their activities more pronounced during the winter season when corpse decomposition is slowed. Disarticulated remains from such activity may be scattered 100 m or more from their original location. Large scavengers tend to devour parts of the corpse in a characteristic sequence beginning with the torso and viscera (Haglund et al. 1989). They will drag parts to woods' edge or nearby ravines for seclusion where they will have screening vegetative cover while feeding. Insects can be transported with these body parts, or they may be attracted first to the scattered remains. It is the complexity of the subcarrion community that will indicate the primary site of corpse decomposition.

Smaller scavengers such as mice, rats, skunks, opossums, raccoons, ravens and crows also will feed on the corpse, but unless the corpse is small, they will not move the remains significantly. They too (except for the birds) are nocturnal and more prevalent in cold seasons. Initial feeding by rodents on flesh can resemble the shallow irregular cratering characteristic of fingernail gouges. On skeletal remains rodent gnawing will be indicated by the marks of their chisel-like incisors. In contrast, insect feeding does not produce artifacts on bone.

Insects such as ants, cockroaches and yellowjacket wasps also may feed on surface tissues and thus create superficial artifacts on the corpse. This feeding produces small, shallow, irregularly shaped craters, which also may be mistaken as fingernail gouges. Also maggots on leaving the corpse may exit by puncturing the thin, somewhat dried skin of the remains. This may give the appearance of shotgun pellet injury. Closer examination should reveal that the rim of the hole will be turned out if the artifact is caused by maggot exiting.

OTHER TYPES OF INSECT INVOLVEMENT

Death scene investigators should be aware that insects and other arthropods also may play an active role in antemortem circumstances. In some cases arthropod activities actually may be the cause of death.

1. **Abuse and Rape -** In death associated with neglectful and/or physical abuse, the kinds of flies present at the remains may

indicate antemortem injury or other circumstances. This also may be true in cases of child or elderly abuse or of rape.

Fecal and urine soiled diapers, garments or bed clothes attract many of the same species of flies as does a decomposing corpse. Both fresh and festering wounds also attract blow flies and flesh flies. Victims who are bound, drugged, intoxicated, helpless or otherwise physically incapacitated may be subjected to fly attack prior to death. Fly maggots of many species will feed on living as well as postmortem tissue. This is termed myiasis. Accurate age determination of the maggots collected will indicate the possibility of antemortem development. Any maggot concentrations at places other than the exposed natural openings of the head should be considered as possible antemortem invasion sites. This is particularly so with bodies in the fresh or bloat stages of decomposition. Maggot concentrations in the urogenital or anal areas suggest traumatic injury or excrement accumulation in that region. Flies do not normally concentrate their oviposition in the urogenital and anal areas unless the site has been predisposed to attract them. Thus, in neglectful child abuse and in rape, the age of maggots at these sites may indicate antemortem invasion. Further, there are a number of fly species which only attack living tissue. Their presence clearly indicates antemortem invasion. Examples of these species are primary screwworms (e.g., *Cochliomyia hominivorax, Chrysomya bezziana*) and certain flesh flies (e.g., *Wohlfahrtia magnifica*).

2. **Bites and Stings -** Deaths caused by scorpion, wasp, bee and ant stings and by spider bites far outnumber those of other venomous animals (Harwood and James 1979). Arthropod venom can cause death directly through its innate potency or indirectly by inducing anaphylactic shock in the victim. Often these attacks could go unnoticed and death may be attributed to heart attack. The arthropod venom is introduced by a break or puncturing of the victim's skin. Generally, there will be a localized swelling and reddening reaction at the site of envenomization, but in the case of acute anaphylaxis there may be little evidence of a local reaction.

   Tick paralysis is yet another form of envenomization which may lead to progressive disability of the victim and death in a matter of hours. In cases where the victim is alone at the time of tick attachment or during the reaction, the cause of death may be inapparent. Generally, the tick will remain attached for sometime after the death of its host.

Details of the medical and pharmacological aspects of arthropod venoms is beyond the scope of this text. The interested reader is encouraged to refer to Tu 1984 (Volume 2) for more complete background on arthropod venoms.

3. **Absence of Insects** - The absence of insects at death scenes where they normally do occur indicates an unusual sequence of postmortem events. This suggests that the remains were held for a period of time in a place inaccessible to insects (e.g., kept submerged, deeply buried, in a freezer, in a sealed trunk, in a tightly closed closet or room, in a screened enclosure, in the trunk of a car). If this has occurred, then the anticipated community of insects and other arthropods will likely be incomplete. However, the total exclusion of insects throughout the entire decomposition succession is only a remote possibility. Insects are extremely adept at invading decomposing remains. More often than not they will succeed.

   If the phase of corpse decomposition does not correspond to the kinds and appropriate age of insects found, then the investigator should suspect that the process has been interrupted or modified in some manner.

CHAPTER 9

THE PAPER TRAIL: CASE RECORDS AND REPORTS

E. P. Catts

Washington State University, Pullman, WA

The machinery of justice in a court-of-law is oiled and fueled by documentation. As is the case in the handling of any evidence in criminal investigations, as well as in the collecting of biological specimens, an accurate record of the source, transfer, and disposition of entomological evidence collected during the course of a death scene investigation is essential.

There are several reasons for this. Firstly, it facilitates the tracking of each piece of evidence even though the item may have been modified following collection. Such modification may include the developmental metamorphosis, or change, of living specimens and dissection, staining or other laboratory manipulation of preserved specimens.

Secondly, there is often a prolonged delay between analysis of evidence and its eventual use in a court-of-law. This calls for a detailed record documenting the transfer and modification of items collected at the death scene at each step in its analysis and storage.

Thirdly, the eventual publication of case studies involving the use of entomological specimens can add to the growing base of information relating to forensic entomology. As with any scientific publication, these case studies must be based on accurate, detailed records.

The purpose of this chapter is to help the death scene investigator, medical examiner or coroner, and entomologist to cooperate in developing a credible paper trail regarding the material collected, processed, and analyzed during the course of an investigation. It may help standardize the content and form of the final report submitted by the forensic entomologist.

For the forensic entomologist, records relating to a death scene investigation may be grouped into eight categories:

1. Initial contact notes
2. Evidence submitted (i.e., photo and video records, and specimen labels)
3. Death Scene Case Study Form (see Appendix A.2)
4. Autopsy report
5. Local weather records
6. Specimen Disposition and Identification Record (see Appendix A.4)
7. Chain of custody record and certified mail receipts
8. Case Study Final Report (see Appendix A.5)

Not all of these documents will be developed in every investigation. However, in general, the more complete the documentation, the more credible the findings will be perceived in court.

The remainder of this chapter will discuss each of the record categories in the order listed above with particular attention given to the format of the Case Study Final Report. In the final analysis, all other documentation serves only to support the final report.

INITIAL CONTACT NOTES: When the forensic entomologist (FE) is first contacted with a request for expertise, written notations should be made regarding certain preliminary information, anticipated services, and fee schedule. Preliminary information includes the name, address, and phone number of the contact-investigator, including the investigating agency being represented. The exact location of the death scene, the phase of decomposition of the corpse, the kind of habitat involved, the date when the deceased was last seen alive (if known), and a general description of the manner of death are appropriate elements of preliminary information to be requested and recorded. However, avoid making any inquiry regarding the estimated postmortem interval made by others so that your own estimate will be made independently and without bias.

Inquiry also should be made regarding the kind and amount of insect activity observed and whether or not the body already has been moved. Often the request for entomological expertise will come at the autopsy well after corpse recovery.

The final item discussed in the Initial Contact should be a clear and mutual understanding as to what expertise the entomolgist will provide and what fee will be charged for this service. The fee charged may vary considerably depending on the category of requesting agency and the reasonable value of time and the cost of support services. Police units or county coroner's offices do not generally have as large an operating budget as do the state or city prosecutor. If the FE visits the death scene the requesting party should also pay any expenses in travel or subsistence incurred. Any time a Case Study Final Report is prepared, a reasonable laboratory service fee is justified, and this should be mutually agreed upon in advance. A reasonable amount might be the equivalent of one day's pay for the FE and additional stipend for laboratory or clerical assistants. Mailing costs and laboratory rearing supplies (containers, rearing medium, etc.) should also be included. Most universities also charge a 40 to 60% overhead for indirect costs. If the FE appears in court as an expert witness, then the additional fee can range from $500 to $1000 or more for each day of involvement in addition to travel and subsistence expenses.

EVIDENCE: If live and preserved specimens have been collected already, then instruction in the best method of handling and transport should be given to the death scene investigator. Although it is best for the FE to do the collecting of specimens at the death scene in person, this is not always possible. Often the death scene has been processed and the corpse moved well before the need for an entomologist is realized. Also, often it is not practical for the entomologist to travel to a death scene some distance away. In such cases the shipment of both live and preserved specimens is required. As a general rule, if live specimens are to be transported for an hour or more, they should be cooled during shipment to arrest development. For shipment, specimens, living or dead and collected as described in Chapter 5, should be chilled (not frozen) and packed in a lightweight insulated container with enough artificial ice coolant (not dry ice) to keep them inactive in transit. This container should be shipped to the entomology laboratory by the most expeditious means available. In some instances the shipment can be by surface transportation (patrol car, public bus, etc.). Over longer distances shipment is by public surface, air or mail routes. The package must be closed with a tamper proof seal to assure delivery intact.

Once items of evidence have been received, they should be stored or processed under conditions of reasonable security that allows only privileged access. This means that live specimens should be kept in a locked incubator or laboratory while being reared. Preserved specimens also should be handled and stored in a similar manner to avoid suspicion in court of pretrial tampering.

Reared specimens should be pinned or preserved in fluid and labeled to indicate the dates of pupation and emergence as well as the case reference number. The proper labeling of evidence was described in Chapter 5 (Also see Appendix A.3).

Any photographs or video tapes made at the death scene should be treated as items of evidence and kept under similar close security. They may be kept in sealed envelopes or a locked filing cabinet and should be marked with the appropriate case reference number.

DEATH SCENE CASE STUDY FORM: This form (Appendix A.2) is a work sheet that itemizes the plethora of conditions to be considered when visiting the death scene. It simply serves as an annotated checklist to minimize the overlooking of potential evidence at the scene. Notations made at the scene on this form will document the source of your information and may suggest circumstances critical to the analysis of samples submitted to the entomology lab.

Much of this information will be duplicated in the investigator's notebook and death scene sketch. Notations regarding the surrounding topography, habitat, temperatures, degree of sun or shade, and amount and kind of insect activity are important to entomological analysis. The back of this form also can be used to record the specific site of collection of each labeled sample and the time of collection. A sample entry might read, "Large, live maggots from left orbit (1420 h)."

The completed form should be kept with the appropriate case file in the entomology laboratory.

AUTOPSY REPORT: The autopsy report should be requested from the medical examiner (ME) or coroner who performs the autopsy. It provides information regarding the **manner** and **cause** of death and the location and approximate extent of insect involvement in corpse decomposition. The ME also may be the first person to notice and collect insect specimens in the investigation. The kind of insect activity and the specific site of their occurrence also will be noted. This information may help the forensic entomologist to estimate the possible influence of maggot-mass generated heat and duration of insect involvement in corpse decomposition.

LOCAL WEATHER RECORDS: Accurate local weather records are an important element in any death investigation involving insects. After a corpse has cooled to ambient conditions, the rate of early decomposition fluctuates to a greater degree with environmental conditions. Weather conditions can vary widely in a relatively small geographic area (.5 mi. radius or less), so local weather records should be used judiciously in estimating insect growth rates which are subject more to microenvironmental conditions. Ideally, a death scene record of shaded air and ground surface temperatures should be made for several days following removal of the corpse in order to determine any consistent differences of conditions at the death scene with those of the weather station record (see Chapter 5). The authenticity of weather records should be verified through the contact investigator.

SPECIMEN DISPOSITION AND IDENTIFICATION RECORD: After preserved specimens collected at the death scene or on autopsy arrive at the entomology laboratory they will be processed for identification and possible dissection. Living specimens are generally held for rearing as described in Chapter 7 so that confirming identifications of immature stages may be made with adult specimens. This should be done whenever possible because species identification using immatures often is not reliable nor possible. If specimens do change in appearance in any way through dissection or metamorphosis, these changes should be recorded with reference to the specimen sample number (Appendix A.4). Dissection of specimens often is necessary to examine spiracular or internal skeletal structures of maggots. Following dissection, all pieces of the specimen should be placed together in a separate vial and labeled.

CHAIN OF CUSTODY RECEIPT: Many large investigating agencies require a Chain of Custody form to accompany items of evidence which are passed from depository to laboratory for analysis. This is true also for entomological evidence, and the completed form always should accompany the items of evidence in exchange. The forensic entomologist should keep a copy of this form to verify what items were received or sent. In addition, all items of evidence which are returned to the principal investigator should be returned by certified mail so that the signed receipt of that transaction will document the transfer.

CASE STUDY FINAL REPORT: The most important document in the paper trail is the Case Study Final Report (See Appendix A.5 for an example of a Case Study Final Report). It describes the findings, analysis and conclusions of the forensic entomologist. This report should include the following items:

a. Case identification number (repeated also on all additional pages of the report).
b. Name, address and phone number of the principal contact person.
c. An abstract scenario of the death scene and corpse condition, and an indication of the date and hour when the deceased was last seen alive, if this is known.
d. A list of the evidence on which the conclusion is based.
e. The identification of the specimens submitted and a brief description of rearing procedures if appropriate.
f. An analysis of the faunal specimens leading to an estimate of the PMI and any other notable aspects. This is the Findings and Discussion section of the report.
g. A summarized conclusion based on the findings.
h. The disposition of evidence and of laboratory records developed during the case study.
i. The name of the forensic entomologist making the study and the date of the report.

As in any scientific writing, careful wording with reasonable qualification of statements is necessary because your report may be used in subsequent courtroom proceedings. As an expert consultant in the case, the forensic entomologist must remain completely objective in drawing conclusions without siding for or against any suspects involved. At times other interpretations of the evidence may be plausible, and to be objective, such alternatives should be identified and the probability of their occurrence indicated.

Note, too, that the case number, contact, evidence submitted, and its disposition, conclusion and identity of the forensic entomologist should all appear on the first page of the Final Report. Additional pages are used to explain the procedures used and to discuss the findings and analysis that led to the conclusions appearing on the first page.

The original copy of the report goes to the contact person, and a courtesy copy should be sent to the ME performing the autopsy. One reference copy is kept for your own files. If Death Scene Case

Study and Specimen Disposition and Identification Record Forms (Appendices A.2 and A.4) were developed during your investigation, copies of these should be submitted as an addendum to the Final Report.

Remember that an accurate, clearly marked paper trail documenting your forensic investigation will add greatly to your own credibility and professional stature.

CHAPTER 10

BEING AN EXPERT WITNESS: TESTIMONIAL GUIDELINES

Bernard Greenberg

University of Illinois at Chicago, Chicago, IL

Occasionally, the forensic entomologist may be called upon (subpoenaed or hired) as an expert witness to testify in a court of law. The basis for this testimony is either to support or refute the Case Study Final Report. Such an event immerses the forensic entomologist in the adversary scenario of our legal system. You must be prepared for this transition and keep your scientific objectivity while contributing to the carrying out of justice.

PREPARATION FOR COURT

The Case Study Final Report contains the entomological findings and conclusions prepared for either the prosecution or the defense. In either case, the report may be questioned in the course of the trial, and the forensic entomologist may be called to testify as an expert witness. Therefore, preparation for court begins with your initial contact, proceeds through the gathering of evidence, and hinges on the accuracy of observations, calculations and conclusions. The Final Report should be couched in words chosen with the same care as for scientific publication. Words like 'probably' and 'may be' are appropriate to reflect various degrees of uncertainty. As biologists we deal with levels of uncertainty that are greater in living systems than in chemistry and physics. In any event, we are called as experts to render opinions, not dogma. Inevitably as forensic entomology is more widely recognized and used in criminal investigations, and where the entomological evidence is particularly crucial, entomologists will be used by both sides to confront each other's interpretations in court. Suddenly, one is faced by the opposing attorney whose job is to discredit your report and expert testimony in a crowded courtroom with reporters leaning foward and TV cameras grinding away. The professional reputation so carefully nurtured is suddenly on the line. But a carefully constructed report, reasonably presented, usually prevails (see Chapter 9).

The report should be thoroughly detailed as to the following: date and time material received; how and by whom delivered; inventory of material; what you did with the material and when; how living specimens were prepared for rearing; rearing conditions and date/time of first eclosion(s); method of preparation of preserved specimens for study (e.g., cleared in 10% KOH, lactophenol, etc.); sample size studied as a basis for conclusions; and confirmation of larval or pupal diagnosis from reared adults. If your identifications were confirmed by others, be sure to include their names and affiliations. In only a few cases I have been called to the death scene to do my own collecting. Usually timing and other circumstances do not allow this very desirable initial participation. If you have collected at the scene, be sure to say so. Ask to be taken to the scene if you have not been there and are going to testify.

Law enforcement personnel who contact you are usually ready to volunteer their estimates of the time of death. This may emerge in the initial interview as you elicit circumstances of the case. Sometimes, it is in the forwarding letter that accompanies the specimens. Your early knowledge of this estimate may be perceived as prejudicial. This may be used by the opposing lawyer to cast doubt on your objectivity. It is better to proceed with your estimates and conclusions without this outside guesswork.

Your estimate of time of death (PMI) should include prioritized alternative conclusions, e.g., most probably the murder occurred during the night of __________ and the eggs were laid the following morning; or, it is less probable that the body was in place on the afternoon of ________ and eggs were laid before dark. Give your reasoning, with the strengths and weaknesses of alternative conclusions.

Remember, you are called to provide an estimate and not an absolute truth. Therefore, do not try to be more precise than the evidence allows — resist crossing the subtle line from science to intuition and advocacy. Remain as objective as possible. Johnston and Villeneuve (1897) considered Megnin's invariably correct conclusions uncanny, but they cautioned: "The chief danger to be feared from Megnin's imitators is that they might tend to indulge in guesses having no very solid basis to apply rules to countries and climates where they are inapplicable." They add that "experiments with animal bodies, other than human, are apt to be misleading. . ." Our science is one hundred years old but still young, and strict adherence to a self-critical objective attitude will foster its development.

BEFORE TRIAL

Remember that your Final Report is available for examination by both sides in a court. The prosecution, in a spirit of cooperation, will often allow the defense attorneys an opportunity to interview you as well. You may feel that this puts you at a disadvantage or, at the very least, provides opposing counsel a preview of how you think and react, your competence, and the kind of information you will present in court. It also allows the defense time to seek entomological experts of its own and to study the literature you may have suggested. Following such a meeting, the opposing counsel may decide to accept your report and not to confront your testimony, but you can not count on it. Once, when I served as a witness for the prosecution, I was visited in my office by the Public Defenders. I discussed the relevant fly life cycles, showed them specimens of eggs, larvae and pupae, provided pertinent literature citations (even photocopied some papers for their use), and gave them names of other forensic entomologists. Another time, when working for the defense, attorneys for both sides retired to a room during recess where they previewed the slides I would show in court.

IN COURT

As trial witness, specifically an expert witness, you will usually receive a subpoena to appear at a specific time, date, and courtroom. Your attire should be neat and neutral. When you arrive, you may have to idle away your time in a witness room waiting your turn to be called. After you are sworn in, the summoning counsel will ask your name, occupation, degrees, etc. to establish your credentials and qualify you as an expert witness. To accomplish this you may be led through a question and answer procedure or you may be given free rein to relate your professional training and experience. This is not the time to be modest. If you have 100 publications, 3 books, 12 past grants and 2 present ones, state these matter of factly. Mention seminars presented at other institutions, congresses, fellowships and other special honors and recognition. If you forget to mention a key item, perhaps your attorney will pick it up as he listens while glancing over your vitae. The opposing counsel generally does not challenge, and the judge then makes the decision to admit you as an expert witness. With this done, you will be led through the relevant entomological evidence. Counsel for both sides will be relying primarily on your Final Report as a basis for their interrogation. Whether you are answering from the witness box or giving a slide or chalkboard presentation, always make eye contact with the jurors. Your goal is to inform and educate them with a minimum of untranslated jargon and to convince them of your competence and sincerity. Because insects **are fascinating,** your task is easy, but what is elementary entomology to you will undoubtedly be arcane to the jurors and probably the judge. Often they are not aware that maggots come from eggs laid only by female flies; the entire life cycle is a jumble to them. You should leave the jury convinced that you know much more than you have related, but do so without a superior or arrogant manner. I cannot emphasize strongly enough the need to first educate your attorney in the relevant fundamentals of entomology. I have seen arguments slip away and a case lost where the attorneys failed to ask the right questions because they have not grasped the concept of ADH (accumulated degree hours) in maggot development.

A slide presentation is an excellent educational vehicle and probably a desirable digression for the jurors. Show slides of adults of the specimens involved and others of eggs, larvae and pupae. If the postfeeding larva is the critical stage, describe this wanderer stage, with the tell-tale shrinking of the crop and accumulation of fat. Discuss, if relevant, the habits of the adult: rural vs. urban, flight and oviposition habits, etc.; and for larvae: temperature-dependent rate of development and the effect of maggot mass generated heat. Relate how disturbance of the corpse may affect successional decomposition and fauna, if appropriate.

A slide presentation may require prior court approval. I recently entered a courtroom prepared to show slides. The equipment was ready, but the presentation was not allowed because the defense had not been forewarned of the presentation.

Make your delivery deliberate, precise and with appropriate emphasis. Take cues from your attorney as to how expansive he/she wants your responses to be. As an expert witness you have considerable latitude in giving your answer, but the more you rattle on, the more vulnerable your testimony is to cross-examination. In the cross-examination expect a tight rein, although this is not invariably so. It is a relief when the opposing attorney allows you to expand your answers and explain fully, compared to a frustrating 'yes,' 'no' or 'I don't know.' However, as an expert witness you are not required to hold to such limited responses. If in doubt, ask the judge for permisison to expand your reply. Exceeding the imposed strictures too often can call forth the wrath of the judge — he/she may excuse the jury and give you a reprimand and a warning. This is a difficult adjustment for scientists who see the natural world as complex, conditional and interrelated. A simple 'yes,' or 'no,' requires saintly restraint. I take the advice of a lawyer with 50 years of trial experience — just answer as the attorney requests; your own attorney is there to rehabilitate your testimony on redirect examination.

Often the witness is asked to testify regarding photographs which were probably previously studied. In some courts you may be cautioned to hold the photographs in such a way that the jurors cannot see them and possibly be prejudiced by what they saw. Ask the judge or the questioning attorney before you flash the photos.

Finally, do not be led down the primrose path to answer questions outside your area of expertise. In a murder trial in South Dakota, I was asked to examine photographs of the victim. "Don't those hands look mummified?" the opposing prosecutor demanded. I knew they looked mummified, but I answered, "I don't know." I might have said, "They do to me, but that is not my field." Had I responded 'yes' or 'no,' I would have been caught parading as a pathologist, and this could have thrown a cloud over my rightful qualifications. Do not hesitate to say, "I don't know," if that is the case.

Before leaving the witness stand, you will be instructed as to whether you are dismissed or excused subject to recall. In either case, it is inappropriate for you to dicuss your own testimony with anyone except the judiciary involved once you have left the stand. Other witnesses may be waiting to testify, and your comments could modify their testimony.

FURTHER READING

Cantor, B. J. 1987. The Expert Witness. Civil Evidence Photogr. Seminars, Belmont, MA. 45 p.

Poynter, D. F. 1987. The Expert Witness Handbook: Tips and Techniques for the Litigation Consultant. Para Pub. Co., Santa Barbara, CA. 243 p.

CHAPTER 11

REPORT/WITNESS ETHICS

E. P. Catts

Washington State University, Pullman, WA

The seeking of truth in the judicial sense is not the same as truth seeking in science. Our judicial system is one of adversary positions, prosecution versus defense. Most death scene faunal evidence originates with one side — police agencies or forensic/crime laboratories. Eighty percent of crime laboratory units are part of police agencies, and fully 30% of their personnel are police officers as well (Saks 1989). The forensic scientists in these agencies are ever aware of the dilemma which they face regarding the holding of an unbiased, objective stance in resolving adversary issues. Few forensic entomologists are, or probably ever will be, members of police agencies or police officers. This helps to dispel any preconceived direct responsibility to one adversary over the other. The ethical issue of strict objectivity and unbaised analysis of data must be as paramount to the forensic entomologist as it is for any other forensic scientist.

Seven ethical problem areas of concern to the forensic scientist were identified by Peterson and Murdock (1989). These problems were as follows: (1) misrepresentation of credentials or area of expertise, (2) deliberate falsification of reports or opinions, (3) incompetent or inappropriate use of methods in analysis, (4) drawing conclusions not supported by data, (5) biased or nonobjective presentation of findings or opinions, (6) failing to point out strengths and weaknesses of findings or opinions, and (7) failing to report unethical practices by other scientists. It is far too easy for the entomologist to become emotionally biased, albeit unconsciously so, in investigating homicides of a particularly gruesome or perverse nature. As a forensic scientist, the entomologist must remain completely objective in arriving at conclusions. It is not the role of the entomologist **to decide the guilt or innocence** of a suspect.

It is our role to give as accurate and detailed a picture as possible of events and circumstances surrounding the death.

There are three areas where ethical procedures are particularly critical: (1) in the gathering (sampling) and handling of material evidence, (2) in the wording and pretrial disclosure of the Case Study Report, and (3) in testimony as a court witness (also see Chapter 10).

SAMPLING

Most faunal evidence comes from one of the judicial adversaries — police agencies or crime laboratories. The forensic entomologist must insist on adequate sampling by these agencies to assure an unbiased conclusion. Inept collecting and handling of faunal evidence is usually attributable to a lack of training or experience by death scene technicians. Because the forensic entomologist seldom visits the death scene, an important activity is the instruction of those people who are charged with working the scene. Thoroughness is the single most critical element in collecting evidence either at the death scene or the autopsy. You can refer to Chapter 5 and 6 for details regarding proper collecting techniques. Even though forensic entomologists cannot prevent inept practices in the gathering of faunal evidence, they can (1) provide necessary training for better sampling, (2) insist on thorough and appropriate sampling and (3) reject improper or biased requests on dealing with samples (Lucas 1989).

HANDLING EVIDENCE

Techniques used in handling and rearing of submitted specimens should follow those generally practiced in making reared associations of any insects. In addition, a routine log or record of the rearing medium, conditions and procedures used and the timing of biological events (e.g., pupariation, adult emergence) should be kept. As with other evidence cataloguing, the proper labeling of insect specimens at each step in the handling sequence is important. As a general rule, all specimens should be returned to the investigating agency. However, where long series of specimens are available, the entomologist may keep a few as voucher specimens to avoid questions regarding species identifications later. Retaining specimens is permissible as long as this fact is recorded in the Case Study Report.

REPORTING

The Case Study Report is the most important document in the entomological investigation of a death. As was described in Chapter 9, the Case Study Report should recount the methods used for study of the evidence as well as the results, a discussion of the analysis and clear, carefully worded conclusions.

As stated by Marling,

> "In criminal cases where the standard proof is beyond reasonable doubt, it is highly desirable that complex scientific evidence . . . be so carefully prepared and expressed that the necessity . . . to challenge it is reduced as much as possible" (Marling 1987, in Lucas 1989)

Therefore, include alternative interpretations when appropriate and indicate the probability of the occurrence of each.

The forensic entomologist deals with circumstantial evidence, not direct evidence. Conclusions are based on the diversity and estimated age of faunal evidence colored by perceived environmental influences and the experience of the scientist. Conflicting interpretations will occur, but if objectivity is kept paramount, they should be uncommon.

Pretrial disclosure of your Case Study Report or of your conflicting conclusion should be made available to the opposing side so that their expert can examine your methods and analysis. Encourage the open and complete communication of your findings and conclusions just as you would in any scientific publication. The American Bar Association Standards state that "it is virtually impossible for evidence or information of this kind to be distorted or misused because of its advanced disclosure" (Giannelli 1989).

COURT WITNESS TESTIMONY

Regardless of which adversary enlists the courtroom testimony of the forensic scientist, the testimony should be presented in a non-biased and objective fashion. In those cases where opposing expert witnesses are called to testify, it is unethical and scientifically dishonest for either witness to challenge opposing findings with generalities, part-truths or scientific jargon. Such testimony might confuse or mislead the jury. The role of the expert is to educate and clarify evidence for the judiciary and jury, not to befuddle.

In summary, ethical responsibility lies with you as the forensic scientist. As a guide there are several established principles to follow (see Peterson and Murdock 1989):

1. Maintain honesty with respect to your qualifications, and always stay only within your own area of expertise.

2. Use analytical methods of proven reliability, and be openly communicative of your findings and analysis.

3. Be objective, non partisan and scientifically honest in drawing conclusions.

4. Introduce understandable, balanced (pro and con) reports or testimony into the legal system.

In addition, avoid accepting an excessive fee for your testimony, or a fee with the amount contingent upon the outcome of the case. Your fee should be scaled to reimburse for your time and expertise only. Nothing more.

FLIES, MAGGOTS AND BEETLES STAND READY AND ABLE TO ASSIST IN DEATH INVESTIGATION

GLOSSARY OF COMMON TERMS IN FORENSIC ENTOMOLOGY

NOTE: Most taxonomic names and their common name counterparts are given in chapter 3 and will not be repeated here.

abdomen - The most posterior of the main body regions in arthropods.

abrasion - A skin injury caused by scraping off of superficial skin due to friction against a rough surface.

active decay - That phase of corpse decomposition that follows bloat, is characterized by much maggot activity, and terminates with a rapid decrease in body weight.

ADD - Abbreviation for accumulated degree days.

ADH - Abbreviation for accumulated degree hours.

adipocere - A peculiar waxy substance consisting of salts and fatty acids and formed from the decomposition of corpse tissues, especially in moist habitats.

adult - The reproductive stage of an animal.

advanced decay - That phase of corpse decomposition that follows active decay and is characterized by beetle activity.

algor mortis - The decrease of corpse body heat to ambient temperature following death.

ambient temperature - The fluctuating levels of heat in air.

ametabolous - Growth without metamorphosis.

ant - An insect in the family Formicidae (order - Hymenoptera) which preys on carrion feeding athropods.

antemortem - Prior to death.

antenna - (pl. antennae) - One of the paired segmented sensory appendages located on the head of arthropods above the mouthparts.

aristate - Antenna having an arista, an enlarged bristle on the distal antennal segment.

arthropod - Any of the invertebrate animals with jointed appendages, an exoskeleton consisting of chitin and protein, and an open circulatory system.

artifact - Here any artifically produced change in appearance on a corpse.
autopsy - A physical examination of the corpse through dissection to determine cause of death (also necropsy and postmortem).
bacteria - Microscopic organisms in the kingdom Prokaryota.
beetle - An insect in the order Coleoptera, the largest of all insect orders with over 300 thousand species.
Berlese funnel - A device used to extract arthropods from soil or litter (see Ch. 7).
bite marks - The impressions left by teeth in an object (e.g., skin).
bloat - That transient phase in corpse decomposition that follows the fresh phase and is characterized by excessive swelling, produced by gases trapped internally.
blood group - Identification of individuals by means of the antigenic character of their blood into categories designated as Group A, Group B, Group .AB, Group O.
blood spatter - Scattered spots of blood.
blow fly - A higher fly in the family Calliphoridae also known as bottle flies.
body bag - A heavy waterproof bag, usually closed with a zipper and used to transport a corpse.
bruise - (see contusion)
C - Abbreviation for centigrade or celcius degrees as a measure of heat.
caddisfly - An insect, aquatic as an immature, in the order Trichoptera; not a true fly.
carrion - Decaying animal flesh.
caudal - Referring to the posterior or tail end.
cephalic - Referring to the anterior or head end.
cephalothorax - The anterior body region in some arthropods consisting of the fused head and thorax.
cerci - (sng. cercus) The paired terminal sensory appendages found in some insects.
chelicerate - Possessing chelicerae as principal mouthparts and lacking mandibles.
chitin - A nitrogenous polysaccharide found in the exoskeleton of all arthropods.
chop wound - A deep, gaping type of injury caused by a sharp, heavy object.
circumstantial evidence - That evidence which only suggests an association with a past occurrence.
class - A taxon in classification between phylum and order.
climate - The long-term weather pattern for a given area.
cloud cover - Condensed water vapor suspended in the atmosphere and obscuring the sun.
coarctate pupa - That object type of pupa that develops inside of the hardened, last instar, larval skin.

cocoon - A protective silken envelope.

community - Used here in an ecological sense meaning the interacting populations of organisms in a given area.

compound eye - An eye composed of closely opposed multiple light sensitive organ units call omatidia.

contusion - A bruise that is either superficial or internal. An injury caused by blunt object impact without laceration with surface discoloration due to subsurface hemorrhaging.

copulatory organs - The caudal structures of male and female insects used in mating.

coroner - An officer responsible for determining manner and cause of death.

corpse - A dead human body sometimes referred to as remains or the deceased.

crop - An internal dilated sac of the insect's foregut which receives and stores food prior to digestion in the midgut of the alimentary tract.

crustacean - That group of arthropods represented by crabs, crayfish, lobsters and sowbugs.

damselfly - An insect, aquatic as an immature, in the order Odonata; not a true fly.

death - The loss of life characterized clinically by combined failure of respiration, cardiovascular and nervous system activity.

death scene - The site where a dead body is discovered.

decomposition - Postmortem degenerative rotting of the corpse.

degree day - Total degrees of heat above a given temperature threshold in one 24 h period.

degree hour - Total degrees of heat above a given temperature threshold in one 60 min period.

dew - Condensed moisture on cool surfaces.

disarticulation - The separation of a body part usually at a joint or suture.

disappearing evidence - That evidence which is transitory and subject to rapid change.

distal - Further away from the main body or point of origin.

dobsonfly - An insect in the order Neuroptera, family Corydalidae, which is aquatic as an immature; not a true fly.

dorsal - The upper part of an insect or the back of a human body.

dragonfly - An insect, aquatic as an immature in the order Odonata; not a true fly.

dried remains - The final phase of corpse decomposition that follows advanced decay and is characterized by little faunal activity.

eclosion - The process of hatching from the egg or of emerging as an adult.

ectoparasite - A parasite that lives on the outside surface of its host.

edema - The swelling of tissues due to excess accumulation of fluids (water logging of tissues).

egg - The characteristic reproductive unit of an adult female. Most carrion insects deposit eggs, but a few deposit first instar larvae.

elytra - The hardened protective first pair of wings among beetles.

emergence - The process of leaving the pupa by the new adult insect.

epidermis - The outermost layer of skin.

ethics - A standard of moral behavior.

ETOH - Abbreviation of ethyl alcohol, ethanol (also EtOH)

exarate pupa - That pupa with the developing legs, antennae and wings not adhering to the body of the developing adult insect.

exsanguination - Loss of blood from the circulatory system.

exoskeleton - The outer arthropod body wall.

expert witness - One who testifies in a court of law on matters pertaining to their field of skill or training.

expertness - Having skill or training in a particular field.

FE - Abbreviation for forensic entomologist.

family - A taxon in classification between order and genus which contains a group of related genera.

fauna - Any organism in the animal kingdom.

feces - Animal solid waste material discharged from the rectum through the anus.

filiform - Hairlike or threadlike.

filter-feed - Among aquatic insects, those which sieve food particles from habitat water with comb-like mouthparts.

flesh fly - A higher fly in the family Sacrophagidae.

forensic - Pertaining to the law.

forensic anthropologist - A scientist who studies the physical characteristics of humans in a legal reference.

forensic entomology - The study of insects and related arthropods from a legal aspect.

forensic pathology - The study of a diseased condition from a legal aspect.

forensic toxicologist(ogy) - A scientist who studies poisons in a legal reference.

frass - Beetle feces.

furcula - The terminal, lever structure in springtails (Collembola) used for springing.

garrotting - Asphyxiation caused by a ligature around the neck.

genitalia - Those external sexual structures used in copulation.

genus - A group of closely related species of organisms. The genus is given as the first part of a scientific name.

halteres - The small, rodlike, modified hind wing in flies (Diptera).

head - The anterior body region; in insects it bears the mouth parts, antennae, eyes and ocelli (when present).

hematoma - Accumulation of blood in tissue due to internal hemorrhaging.
hemorrhage - Escape of whole blood from a blood vessel.
holometabolous - That type of metamorphosis where the young and adults are completely different and make their living differently.
homicide - The deliberate killing of one human being by another.
hypothesis - An unproved theory.
incision - A wound produced by a sharp edged instrument or object.
indigenous - Native to a given area.
insect - Any arthropod characterized by having mandibles and six legs at some stage in its life.
instar - Each of the successive incremental growth steps terminated by a molt.
invertebrate - Any species of animal lacking a back bone.
KAAD - See Ch. 7.
Kahle's - See Ch. 7.
KOH - Abbreviation for potassium hydroxide (see Ch. 7).
laceration - A wound produced by a tear in the skin due to application of blunt force in crushing or shearing.
lactophenol - A clearing agent solution (see Ch. 7).
larvae - The primary feeding and growth stage of invertebrates. Usually it is the stage following hatching of the egg.
litter - The surface humus layer of partly decomposed plant material.
livor moris - The gradual discoloration of skin surfaces due to gravitational settling of blood in the corpse.
MD - Abbreviation for Doctor of Medicine.
ME - Abbreviation for Medical Examiner.
MS - Abbreviation for Master of Science.
macro lens - In photography a lens having the capability of .5 to 1.0 magnification (i.e., life size).
maggot - The larva of a higher fly. It sheds its skin twice and has three growth instars prior to pupariation.
maggot-mass - The collective, closely packed mass of higher fly larvae occurring in decomposing carrion.
mandibulate - Possessing mandibles as principle mouthparts and lacking chelicerae.
manslaughter - Homicide without malice.
mayfly - An insect, aquatic as an immature in the order Ephemeroptera; not a true fly.
medical entomology - The study of arthropods that adversely affect the health of vertebrate animals including people.
medicolegal entomology - Medical entomology from a legal aspect.
mesothorax - The middle segment of the thorax.
metamorphosis - The process in arthropods of changing from one life stage into another.
methathorax - The third, or hindmost, segment of the thorax.

microclimate - The long-term weather pattern in a very small area.
midge - A lower fly in the family Chironomidae.
migrate - As used here this means to move away from the corpse, usually used with reference to maggots. Technically this is emigration.
mite - Any arthropod in the order Acari. These are very small to minute animals having four pairs of legs in the adult stage, but only three pairs in the larva. All mites have chelicerated mouth parts and lack mandibles.
molt - The process of shedding of the exoskeleton of outerbody wall in arthropod growth - also termed ecdysis.
moribund - Dying.
mouth hooks - The paired maxillary oral structures of a maggot.
mouthparts - The oral structures of the head in arthropods that function to process and ingest food.
mummification - The drying, shrinking and hardening of dead flesh due to extreme dehydration.
muscoid - Flies closely related to the family Muscidae (house flies).
myiasis - The invasion of any living vertebrate animal, including people, by fly larvae, especially maggots.
necrophilous - Having a dietary fondness for dead flesh.
necropsy - (see autopsy).
neglectful abuse - Injury or mistreatment by not nurturing or caring for a more or less helpless person.
nit - The egg of a louse.
ocellus (pl. ocelli) - A simple isolated light sensitive organ among arthropods.
omnivorous - A tendency to eat anything and everything.
order - A taxon in classification between class and family which includes related familes of organisms.
oviposition - The act of depositing eggs.
paurometabolous - That type of metamorphosis where the young and adults look similar and live in the same habitat; the adults are winged.
PhD - Abbreviation for Doctor of Philosophy.
phylum - A taxon in classification between kingdom and class.
physical abuse - Injury or mistreatment by a deliberately inflicted physical means.
physical evidence - Any tangible article which tends to prove or disprove a point in question.
PMI - Abbreviation for postmortem interval.
postfeeding larva - The wandering, fasting phase of the third instar - maggot, terminating in pupariation.
postmortem - After death.
postmortem interval - The period of time between death and corpse discovery.
precipitation - Rain, snow, sleet, or hail.

predatory - That life style of killing and eating prey.

primary crime scene - The place where the corpse is discovered and homicide or suicide is suspected.

pronotum - The dorsal (upper) portion of the first thoracic segment in insects.

prothorax - The first, or anterior, segment of the thorax.

proximal - Closer or nearer to the main body or point of origin.

psychrometer - A device consisting of wet and dry bulb thermometers used to measure relative humidity.

pupa - That immature stage between the larva and adult in insects having complete metamorphosis. This is a stage of major transformation. Among higher flies the pupa is inside the puparium.

pupariation - The immobilization of the postfeeding maggot with the shrinking, hardening and darkening of its outer skin.

puparium - The hardened outer shell or skin of the final larval instar within which the pupa of a higher fly develops.

pupation - To change form into a pupa in the process of transforming from the larval stage to the adult.

putrifaction - The foul smelling, anaerobic decomposition of moist or wet organic matter by micro-organisms.

rape - The crime of sexual assault on another person without their consent.

relative humidity - The level of moisture vapor in air at a given temperature.

remains - See corpse.

retenaculum - The ventral, trigger structure in springtails (Collembola) used to hold the furcula in the "cocked" position.

RH - Abbreviation for relative humidity.

rigor mortis - The progressive rigidity of a corpse following death caused by an accumulation of lactic acid in dying muscle tissues. This is a temporary condition lasting 12 to 36 hours.

saponification - The conversion of corpse body fat into a curd-like foul smelling product - adipocere.

scanning electron microscopy - A process of photographing very small, metal-coated, objects in extreme, minute detail.

scavengers - Here this refers to any animal which feeds on carrion or other decaying organic matter.

sclerotized - The hardened or tanned part of the insect's cuticle or exoskeleton.

secondary crime scene - Any of a number of sites that relate to the primary crime scene.

secretor - That 8% of the human population whose body secretions can be analyzed for blood type.

seepage - The fluid materials draining from the decomposing corpse.

seral - Adjective form of sere.

sere - One of a series of progressive waves of community change in ecological succession.
soil - The upper layer of undifferentiated surface material in terrestrial ecosystems.
species - A population of organisms that are reproductively isolated from other populations; that is the populations cannot interbreed successfully.
spiracle - The external opening of the arthropod tubular tracheal system through which the animal respires.
stab - A puncturing type of wound where penetration is deeper than the width of the entry site.
stage - Any one of the various forms, within the life of an individual arthropod - i.e., egg, larva (nymph, maggot) pupa, adult.
subpoena - A legal order to appear in a court-of-law.
succession - Used here has ecological succession. The progressive replacement of one ecological community with another.
suicide - The deliberate killing of one's self.
synanthropic - Having an affinity or preference for humans.
systematics - The study of the phylogenetic relationships among organisms.
taxon - (pl. taxa) - A group of related organisms.
taxonomy - The science of describing and classifying organisms into related groups called taxa.
tegmina - The leathery, protective first pair of wings in cockroaches.
thermal units - Units (usually degrees) used to measure heat intensity.
thorax - The middle body region in insects which bears the legs and wings when present. With vertebrate animals, the portion of the body encompassing the rib cage.
tick - An arthropod (order Acari) in the superfamily Ixodoidea that is a blood feeding parasite on vertebrate animals.
trace evidence - Physical evidence in small amounts.
urban entomology - That area of entomology dealing with arthropods in an urban or nonrural setting.
vertebrate - Any animal with a backbone or vertebral column.
ventral - The under part of an insect and the belly or front of a human body.
viscera - The larger, soft interior organs of the lower half of the body.
wasp - Any winged insect in the order Hymenoptera generally possessing a sting. These may be scavengers, predators or parasites of organisms at carrion.
weather - The general atmospheric conditions of a given area at a given time.
XAAD - See Ch. 7.
yellowjacket wasp - A wasp in the family Vespidae associated with carrion.

REFERENCES CITED

Aldrich, J. M. 1916. Sarcophaga and allies. Thomas Say Found., Lafayette, IN. 301 p., illus.

Bergeret, M. 1855. Infanticide, momification du cadavre. Decouverte du cadavre d'un enfant nouveau-ne dans une cheminee ou il setait momifie. Determination de l'epoque de la naissance par la presence de nymphes et de larves d'insectes dans le cadavre et par l'etude de leurs metamorphoses. Ann. Hyg. Med. Leg. 4:442-452.

Bornemissza, G. F. 1957. An analysis of arthropod succession in carrion and the effect of its decomposition on soil fauna. Aust. J. Zool. 5: 1-12.

Borror, D. J., C. A. Triplehorn, and N. F. Johnson. 1989. An introduction to the study of insects, 6th Edn. Sanders College Publishing, Philadelphia, 875 p.

Cornaby, B. W. 1974. Carrion reduction by animals in contrasting habitats. Biotropica 6: 51-63.

Early, M. and M. L. Goff. 1986. Arthropod succession patterns in exposed carrion on the island of Oahu, Hawaiian Islands, USA. J. Med. Entomol. 23: 520-531.

Fox, R. H. and C. L. Cunningham. 1973. Crime scene search and physical evidence handbook. U. S. Dept. of Justice, Washington, D. C. p. 11-43.

Geberth, V. J. 1990. Practical Homicide Investigation. 2nd Ed. Elsevier Sci. Publ. Co., Inc., New York. 599 p.

Giannelli, P. C. 1989. Evidentiary and procedural rules governing expert testimony. J. Forensic Sci. 34: 730-748.

Gilbert, B. M. and W. M. Bass. 1967. Seasonal dating of burials from the presence of fly pupae. Amer. Antiquity 34: 534-535.

Goddard, J., and P. K. Lago. 1983. An annotated list of the Calliphoridae (Diptera) of Mississippi. J. Ga. Entomol. Soc. 18: 481-484.

Goff, M. L. 1989. Gamasid mites as potential indicators of postmortem interval. *In.* Progress in acarology, Vol. 1, G. P. Channabasavanna & C. A. Viraktamaths, eds., Oxford & IBH Publishing Co. PVT, LTD, New Delhi. p. 443-450.

Goff, M. L. and C. B. Odum. 1987. Forensic entomology in the Hawaiian Islands. Three cases. Amer. J. Forensic Med. and Pathol. 8: 45-50.

Goff, M. L., C. B. Odom and M. Early. 1987. Estimation of post-mortem interval by entomological techniques: A case study from Oahu, Hawaii. Bull. Soc. Vector Ecol. 11: 242-246.

Goff, M. L., M. Early, C. B. Odom, and K. Tullis. 1986. A preliminary checklist of arthropods associated with exposed carrion in the Hawaiian Islands. Proc. Hawaii Entomol. Soc. 26: 53-57.

Goff, M. L., A. I. Omori and J. R. Goodbrod. 1989. Effect of cocaine in tissues on the development rate of *Boettcherisca peregrina* (Diptera: Sarcophagidae). J. Med. Entomol. 36: 91-93.

Greenberg, B. 1985. Forensic entomology: case studies. Bull. Entomol. Soc. Amer. 31: 25-28.

Greenberg, B. 1990. Blow fly nocturnal oviposition behavior. J. Medical Entomol. 27: 807-810.

Gunatilake, K. and M. L. Goff. 1989. Detection of organophosphate poisoning in a putrifying body by analyzing arthropod larvae. J. Forensic Sci. 34: 714-716.

Haglund, W. D., D. T. Reay, and D. R. Swindler. 1989. Canid scavenging/disarticulation sequence of human remains in the Pacific Northwest. J. Forensic Sci. 34: 587-606.

Hall, D. G. 1948. The Blowflies of North America. The Thomas Say Foundation, Lafayette, IN. 477 p., illus.

Hall, R. D. 1979. The blow flies of Missouri: An annotated checklist (Diptera: Calliphoridae). Trans. Mo. Acad. Sci. 13: 33-36.

Hall, R. D. and L. H. Townsend, Jr. 1977. The blow flies of Virgina. The Insects of Virginia No. 11, Va. Polytech. Ins. Res. Div. Bull. No. 123, 48 p., illus.

Harwood, R. F. and M. T. James. 1979. Entomology in Human and Animal Health. Macmillan Publ. Co., Inc., New York, Toronto, London. 548 p.

Haskell, N. H., D. B. McShaffrey, D. A. Hawley, R. E. Williams, and J. E. Pless. 1989. Use of aquatic insects in determining submersion intervals. J. Forensic Sci. 34: 622-632.

Hawley, D. A., N. H. Haskell, D. G. McShaffrey, R. E. Williams, and J. E. Pless. 1989. Identification of a red "fiber": Chironomid larvae. J. Forensic Sci. 34: 617-621.

Hoffman, R. L. and J. A. Payne. 1966. Diplopods as carnivores. Ecology 50: 1096-1098.

Holzer, F. J. 1936. Zerstorung an wasserleichen durch larven des kocherfliege. Zeitshrift Fur Die Gesamte Gerichtliche Medizin 31: 223-228.

Jiron, L. F., and V. M. Cartin. 1981. Insect succession in the decomposition of a mammal in Costa Rica. J. N. Y. Entomol. Soc. 89: 158-165.

Johnston, W. and G. Villeneuve. 1897. On the medico-legal application of entomology. Montreal Med. J. 26: 81-90.

Kamal, A. S. 1958. Comparative study of thirteen species of sarcosaprophagous Calliphoridae and Sarcophagidae (Diptera). Ann. Entomol. Soc. Amer. 51: 261-271.

Knipling, E. F. 1936. A comparative study of the first-instar larvae of the genus *Sarcophaga,* with notes on the biology. J. Parasitol. 22: 417-454.

Leclercq, M. 1969. Entomology and legal medicine. *In* Leclercq, M. ed. Entomological Parasitology: The Relations between Entomology and the Medical Sciences. Pergamon Press, Ltd., Oxford. p. 128-142.

Leclercq, M. and L. Quinet. 1949. Application d' entomologie dans la determination de la date de la morte. Ann. Med. Leg. 29: 324.

Liu, D. and B. Greenberg. 1989. Immature stages of some flies of forensic importance. Ann. Entomol. Soc. Amer. 82: 80-93.

Lord, W. D. and J. F. Burger. 1983. Collection and preservation of forensically important entomological materials. J. Forensic Sci. 28: 936-944.

Lord, W. D. and J. R. Stevenson. 1986. Directory of forensic entomologists, 2nd. ed. Def. Pest Mgmt. Info. Anal. Center, Walter Reed Army Medical Center, Washington, D.C. 42 p.

Lord, W. A., E. P. Catts, D. A. Scarboro, and D. B. Hadfield. 1986. The green blow fly, *Lucilia illustris* (Meigen), as an indicator of human post-mortem interval: A case of homicide from Fort Lewis, Washington. Bull. Soc. Vector Ecol. 11: 271-75.

Lucas, D. M. 1989. The ethical responsibilities of the forensic scientist: Exploring the limits. J. Forensic Sci. 24: 719-729.

McCafferty, W. P. 1981. Aquatic entomology. Jones and Bartlett Publishers, Inc. Boston. p. 448.

McKnight, B. E. 1981. The Washing away of Wrongs: Forensic Medicine in Thirteenth-Century China. Ann Arbor: Univ. Mich. 181 p.

Meek, C. L., M. D. Andis, and C. S. Andrews. 1983. Role of the entomologist in forensic pathology, including a selected bibliography, Biblio. Entomol. Soc. Amer. 1:1-10.

Megnin, J. P. 1894. La Faune des cadavres: application e l'entomologie a la medecine legale. Encyclopedie Scientifique des Aide-Memoires, Masson et Gauthiers — Villars, Paris. 214 p.

Merritt, R. W. and K. W. Cummins, (Eds.). 1984. An introduction to the aquatic insects of North Amercia. 2nd Ed. Kendall-Hunt Publishing Co., Dubuque, Iowa, 722 p.

Motter, M. G. 1898. A contribution to the study of the fauna of the grave. A study on one hundred and fifty disinterments with some additional experimental observations. J. N. Y. Entomol. Soc. 6: 201-231.

Nuorteva, P. 1987. Empty puparia of *Phormia terraenovae* (Diptera, Calliphoridae) as forensic indicators. Ann. Entomol. Fenn. 53: 53-56.

Nuorteva, P., M. Isokoski, and K. Laiho. 1967. Studies on the possibilities of using blowflies (Dipt.) as medicolegal indicators in Finland. 1. Report of four indoor cases from the city of Helsinki. Ann. Entomol. Fenn. 33: 217-225.

Nuorteva, P. H., Schumann, M. Isokoski, and K. Laiho. 1974. Studies on the possibilities of using blowflies (Dipt. Calliphoridae) as medicolegal indicators in Finland. 2. Four cases where species identification were performed from larvae. Ann. Entomol. Fenn. 40: 70-74.

Nuorteva, P. 1977. Sarcosaprophagous insects as forensic indicators. *In* Forensic Medicine: A Study of Trauma and Environmental Hazards. Vol. II, W. B. Saunders Co., Philidelphia. pp. 1072-1095.

Payne, J. A. 1965. A summer carrion study of the baby pig *Sus scrofa* Linnaeus. Ecology 46: 592-602.

Payne, J. A. and D. A. Crossley. 1966. Animal species associated with pig carrion. Health Phys. Div. A Prog. Rep. (Oak Ridge National Laboratory) Tennessee, ORNL-TM-1432. 70 p.

Payne, J. A., E. W. King, and G. Beinhart. 1968. Arthropod succession and decomposition of buried pigs. Nature, London. 219: 1180-1181.

Payne, J. A. and E. W. King. 1969. Lepidoptera associated with pig carrion. J. Lepid. Soc. 23: 191-195.

Payne, J. A. and W. R. M. Mason. 1971. Hymenoptera associated with pig carrion. Proc. Entomol. Soc. Wash. 73: 132-141.

Peterson, A. 1967. Larvae of insects, an introduction to nearctic species. 6th Edn., Printed for the author by Edward Brothers, Ann Arbor, Michigan, 315 p.

Peterson, J. L. and J. E. Mudock. 1989. Forensic science ethics: Developing an integrated system of support and enforcement. J. Forensic Sci. 34: 749-762.

Redi, F. 1668. Esperienze intorno alla generazione degl' insetti. Insegna della Stella, Florence. 228 p.

Reed, H. B., Jr. 1958. A study of dog carcass communities in Tennessee with special reference to insects. Amer. Midl. Nat. 59: 213-245.

Rodriguez, W. C. and W. M. Bass. 1983. Insect activity and its relationship to decay rates of human cadavers in east Tennessee. J. Forensic Sci. 28: 423-432.

Rodriguez, W. C. and W. M. Bass. 1985. Decomposition of buried bodies and methods that may aid in their location. J. Forensic Sci. 30: 836-852.

Saks, M. J. 1989. Prevalence and impact of ethical problems in forensic science. J. Forensic Sci. 34: 772-793.

Simpson, K. W. and R. W. Bode. 1980. Common larvae of the Chironomidae (Diptera) from New York state streams and rivers, with particular reference to the fauna of artificial substrates. Bull. New York St. Mus., Bull. No. 439, 1-105.

Smith, K. G. V. 1986. A Manual of Forensic Entomology. Cornell University Press. Ithaca, NY. 205 p.

Tu, A. T. (Ed.) 1984. Handbook of Natural Toxins: (Vol. 2) Insect Poisons Allergens and Other Invertebrate Venoms. Marcel Dekker, Inc., New York and Basel, 732 p.

Tullis, K. and M. L. Goff. 1987. Arthropod succession in exposed carrion in a tropical rainforest on Oahu Island, Hawaii. J. Med. Entomol. 24: 332-339.

Vincent, C. D., K. McE. Kevan, M. Leclercq, and C. L. Meek. 1985. A bibliography of forensic entomology. J. Med. Entomol. 22: 212-219.

Williams, H. 1984. A model for the aging fly larvae in forensic entomology. Forensic Sci. Internat. 25: 191-199.

Appendix A.1

COLLECTING INSECTS FOR HOMICIDE INVESTIGATIONS

Contact:

Dr. I. M. Wright
Dept. Entomology
Blissful State Univ.
Collegeville, USA
(923) 338-7436

Equipment:

1. Hand Net
2. Forceps & Digging Tool
3. Thermometer
4. Vials, Jars, Plastic Bags

Supporting Data Needed

1. Previous weather for area
2. On site weather data (3-5 days)
3. Photos/video of crime scene
4. Clock time of collecting
5. M.E. Autopsy Report (when available)

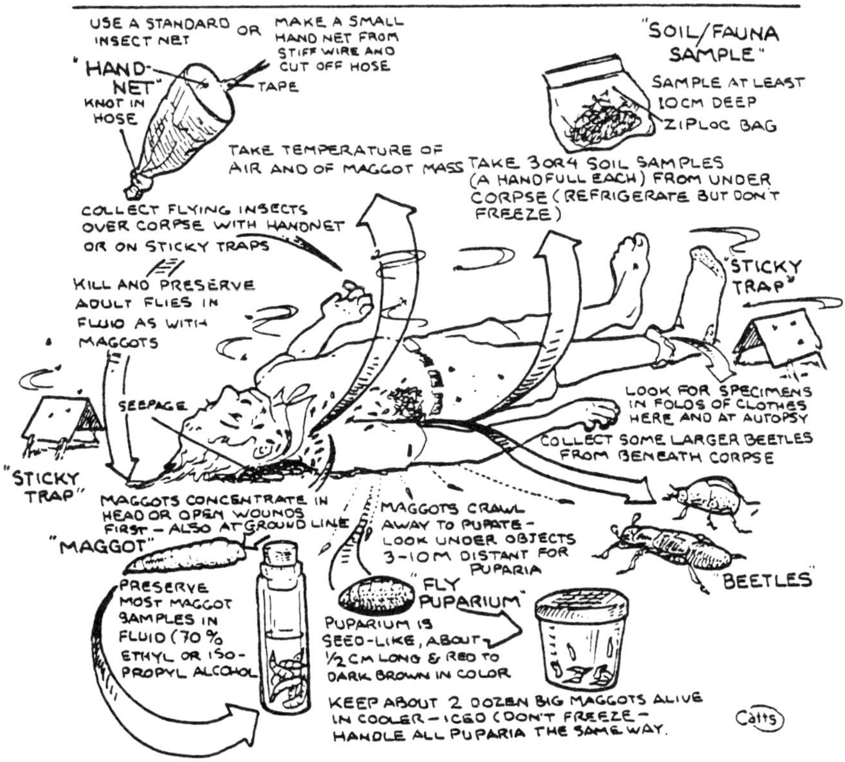

Appendix A.2

DEATH SCENE CASE STUDY FORM

CASE STUDY #________________ DATE ________________________

NAME ________________________ AGE _______ SEX ___________

DATE FOUND ____________ DATE REPORTED MISSING ____________

LAST SEEN ALIVE __________ LOCATION FOUND ____________________
City/County State

INVESTIGATOR ____

SITE DESCRIPTION ____

____

____

DEATH SCENE AREA

RURAL ______________

forest ___ tillable field ___ pasture ___ brush ___ roadside ___

barren area ___ closed building ___ open building ___

other ____

URBAN/SUBURBAN ______________

closed building ___ open building ___ vacant lot ___

pavement ___ trash container ___ other ______________

AQUATIC HABITAT ______________

pond ___ lake ___ creek ___ small river ___ large river ___

irrigation canal ___ marine environment ___ other ________

EXPOSURE Open air ___ burial ___ (depth) other ______________

Comments ____

ENCLOSED ___ entire ___ partial ___ nude ___

portion of body clothed ______________________________

____

STAGE OF DECOMPOSITION fresh ___ bloat ___ active decay ___

advanced decay ___ remains ___ saponification ___ other _____

Comments ____

EVIDENCE OF SCAVENGERS ____________________________________

TRAUMA CONTRIBUTION TO INSECT ACTIVITY ____________________

____

TEMPERATURES AT SCENE ambient _____ body surface _____ ground surface _____ under-body interface ___ soil ___ (10cm, 20cm depth) maggot mass ___ water temperature (if aquatic) ___ enclosed structure ___ (record temperatures (ambient, ground, soil) periodically each day at site for 2 - 5 days after body recovery)

NUMBER OF SAMPLES OBTAINED preserved _____ live _____

WEATHER DATA ____

____

____

Appendix A.3

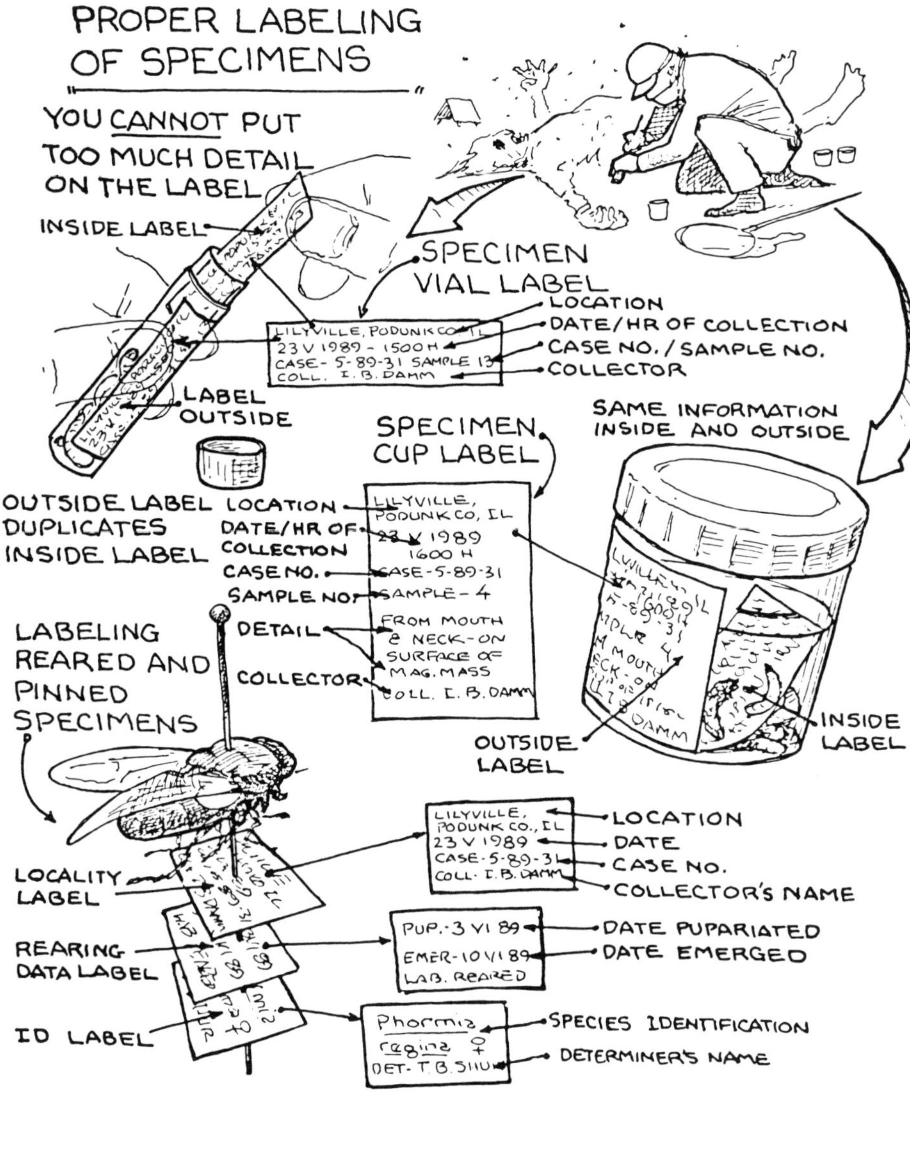
PROPER LABELING OF SPECIMENS
"YOU CANNOT PUT TOO MUCH DETAIL ON THE LABEL"
INSIDE LABEL
SPECIMEN VIAL LABEL
LILYVILLE, PODUNK CO. IL
23 V 1989 - 1500 H
CASE- 5-89-31 SAMPLE 13
COLL. I. B. DAMM
LOCATION
DATE/HR OF COLLECTION
CASE NO. / SAMPLE NO.
COLLECTOR
LABEL OUTSIDE
SAME INFORMATION INSIDE AND OUTSIDE
SPECIMEN CUP LABEL
OUTSIDE LABEL DUPLICATES INSIDE LABEL
LOCATION
DATE/HR OF COLLECTION
CASE NO.
SAMPLE NO.
DETAIL
COLLECTOR
LILYVILLE, PODUNK CO, IL
23 V 1989
1600 H
CASE-5-89-31
SAMPLE-4
FROM MOUTH & NECK-ON SURFACE OF MAG. MASS
COLL. I. B. DAMM
OUTSIDE LABEL
INSIDE LABEL
LABELING REARED AND PINNED SPECIMENS
LOCALITY LABEL
REARING DATA LABEL
ID LABEL
LILYVILLE, PODUNK CO., IL
23 V 1989
CASE-5-89-31
COLL. I. B. DAMM
LOCATION
DATE
CASE NO.
COLLECTOR'S NAME
PUP.-3 VI 89
EMER-10 VI 89
LAB. REARED
DATE PUPARIATED
DATE EMERGED
Phormia regina ♀
DET- T. B. SHU
SPECIES IDENTIFICATION
DETERMINER'S NAME

Appendix A.4

SPECIMEN DISPOSITION AND IDENTIFICATION RECORD

CASE # ______________ COUNTY/STATE ______________

Date Obtained	Sample Number	Location Obtained	Date of Identification	Number of Specimens	Stage/ Instar	Family/Species

IDENTIFIER ______________________

Appendix A.5

Example Forensic Case Study Final Report
(NOT AN ACTUAL CASE)

CASE NO: 88-063325 Ph (923) 338-7436
(homicide, Kathrine Morten)

CONTACT: Detective Richard S. Tracy
Homicide Division
Public Safety Building
Nowhere City, USA Ph (923) 548-3737

EVIDENCE SUBMITTED:

1. Photographs (24 in color, 8 × 10) of death scene and victim.
2. Video tape (25 min) of death scene and corpse during removal.
3. Autopsy report - M.E., Podunk Co.
4. One container of preserved insects, mostly maggots (220+) collected at 1330 hours, 15 May from face and neck of deceased.
5. Two plastic bags of live maggots (about 50 in each bag) collected from upper and lower areas of corpse at autopsy. Live maggots were delivered to this lab. at 1830 hrs. 17 May by Podunk Co. Sheriff's Deputy.
6. Weather data, Nowhere City International Airport, 5 May - 15 May 1988.
7. Deceased - Last seen alive - 0400 hrs, 5 May
   - Corpse discovered - 1015 hrs, 15 May
   - Autopsy performed - 0900 hrs, 17 May

CONCLUSION: The estimated postmortem interval (PMI) from faunal evidence is six (6) days, placing the time of death between 1800 hrs 7 May and 1000 hrs. 8 May. Maggots in the genital/anal area indicate assault prior to or just after death. Continuous rain from the night of 11 May through 13 May did not depress maggot mass development. Blow flies (*Phormia regina*) probably laid eggs on the corpse on mid-afternoon of 9 May. Preserved maggots were determined to be approximately 5.5 days old.

FORENSIC ENTOMOLOGIST: ______________________________
I. M. Wright
Professor, Ph.D.

DATE: 10 June 1988.

Case No. 88-063325 (p. 2)

SCENARIO: The nude body of the victim, a caucasian female of about 20 years of age, was discovered at 1015 hrs., 15 May 1988 in an open pine woodland about 50 ft. off a logging road 8 mi NW of Blissful City. The corpse was positioned face down with the multiple stab punctures on the dorsal, upper thorax. The wrists were bound with electrical cord behind the victim's back. A large amount of dried, pooled blood was under the body. The corpse was in late bloat with an extensive maggot mass beneath the head and upper torso and with a smaller mass in the genital/anal area. Blackening of face and hands had occurred. The body was lying in a small, flat, open area carpeted with about 2 inches of Ponderosa pine needles.

FINDINGS AND DISCUSSION: The live maggots (collected at autopsy) were placed on beef liver on vermiculite and covered with a dampened paper towel. They were held at 75 - 80F and 75 RH until adults emerged 8 days later. The only species reared from maggots in live collections was *Phormia regina*, the black blow fly. This fly typically arrives within the first 24 hours at a fresh corpse and often within 15 min.

The larger preserved maggots were late 3rd instar. Their crops were still distended with blood indicating they were actively feeding. They were of mixed age classes but the larger maggots from both body sites were 14 - 15 mm long indicating an age of about 5 - 6 days. Development of these massed maggots probably was not depressed by the cool, wet weather which covered all of the eastern part of the state on 11 - 13 May. Growth also would not be arrested in the 40 hours preceding autopsy when the bagged body was refrigerated. Weather conditions during the days of 6 - 8 May were unusually warm with broken cloud cover and were ideal for blow fly activity. Air temperatures then dropped below 50F on 9 May prior to the beginning of 3 days of cloud cover and rain.

The size of the maggots and the occurrence of favorable weather on 8 May indicates that blow fly eggs producing the maggots examined were laid on 8 May. If eggs had been laid earlier maggot development would have been further advanced. The estimated PMI then is six (6) days and places the time of death between evening on 7 May and mid-day on 8 May.

The presence of well developed maggot mass in the genital/anal area of the victim also indicates sexual assault immediately before or after death. Blow flies rarely oviposit in the genital/anal area unless trauma and bleeding has occurred at those openings. Other insects present included one large carrion beetle (Siphidae) and 4 ants (Formicidae). Neither of these are reliable indicators of the PMI.

Case No. 88-063325 (p. 3)

DISPOSITION OF EVIDENCE:

1. Reared flies (adults and puparia) are returned under separate cover (certified delivery).
2. Photos are returned with this report.
3. Weather data and video tape copy are retained in this lab. for reference.

INDEX

G

H

I

NOTES

NOTES

NOTES

NOTES

NOTES

NOTES